A   PEREGRINE   SMITH   BOOK

Gibbs M. Smith, Inc.
Salt Lake City

WHARTON ESHERICK   ARTHUR ESPENET CARPENTER

CONTEMPORARY

GEORGE NAKASHIMA   JAMES KRENOV

AMERICAN

BOB STOCKSDALE   WENDELL CASTLE

WOODWORKERS

TAGE FRID   GARRY KNOX BENNETT

MICHAEL A. STONE

SAM MALOOF   JERE OSGOOD

Copyright © 1986 by Michael A. Stone

This is a Peregrine Smith Book

Published by Gibbs M. Smith, Inc.

P.O. Box 667

Layton, UT 84041

Printed and bound in Japan

Printing. 90 89 88 87 86   5 4 3 2 1

Library of Congress Cataloging in Publication Data
Stone, Michael A., 1954-
    Contemporary American woodworkers.
    bibliography: p.
    Includes index.
    1. Woodworking—United States. 2. Furnituremaking—United States.
3. Woodworkers—United States.
I. Title.
TT194.S75 1985 749.213 [B] 85-11797
ISBN 0-87905-098-5

Photo credits: front cover flap—copyright © Steven Sloman, courtesy Alexander F. Milliken, Inc.; back cover flap—copyright © Esto Photographics, Inc.

Frontispiece: a bowl of Philippine ebony by Bob Stocksdale (photo credit: Stone & Steccati)

For Louise

Book design by Clane Graves

TABLE OF CONTENTS

PREFACE

The idea for a book on America's leading wood craftsmen first occurred to me while writing an article on Wharton Esherick for *Fine Woodworking* magazine. Having grown up just six miles from Esherick's studio, I was familiar with his work, especially the furniture he created for the Hedgerow Theatre where I had studied dramatics in high school in hope of becoming a famous actor. During the course of my research for that story, I interviewed Wendell Castle, who told me he first considered a career making furniture after seeing Esherick's work in Don Wallance's book, *Shaping America's Products.* "Seeing those pieces," said Castle, "I suddenly realized the possibilities of combining art and furniture." Wallance's reference to Esherick consisted of just a few paragraphs and three pictures. If that scant passage could inspire the career of an eminent designer-craftsman such as Wendell Castle, I reasoned, a well-illustrated book examining the designs, techniques and philosophies of several notable American woodworkers would surely make a valuable contribution. Soon after my story on Esherick was published in 1979, Gibbs Smith, president of Peregrine Smith Books, contacted me with a proposal to write a book of similar scope.

My plan was straightforward: I would select ten contemporary American woodworkers who had made profoundly significant contributions to the craft. Although many people write with nostalgia of the furniture of yesteryear (in reality, American antiques are often plagiaristic and poorly made), they generally ignore the many fine craftsmen who are working today. But to me, the modern artisans are the most interesting because they use an ancient craft to express contemporary aesthetics. Applying both conventional and innovative techniques, they create furniture and other utilitarian objects that reflect the images and shape the tastes of our times.

I began the process of selecting the craftsmen to include by interviewing over 100 craftsmen, curators, teachers, students, collectors and gallery owners. I asked them who they felt were the most significant woodworkers of our time. Equipped with a list of names mentioned most frequently, I continued my research in the field. Visiting the workshops and homes of woodworkers around the country, I studied their methods and observed their life-styles. In addition to hours of personal interviews, I attended their lectures and classes to watch them relate to students and other craftsmen. I was happily surprised to meet individuals whose work surpassed their modest celebrity, and disappointed to find craftsmen whose designs and techniques fell short of their reputations.

I also traveled to the homes of collectors to view commissions in their intended settings. I visited the former home of Judge Curtis and Nellie Lee Bok, who supported Wharton Esherick during the lean years of the Depression. Esherick, in turn, fashioned the interior of their home into a masterpiece of sculpted wood. The collectors I met were of various means and tastes—one of them, David L. Davies, constructed the inside of his New Jersey estate to accommodate a 16-foot spiral staircase by Arthur Espenet Carpenter. Lynn E. Farkas, a no less astute and passionate collector, devotes much of her income as a clerk to wooden turnings and other objects, which, for lack of space, she stores under her bed.

Eventually, I narrowed my list to 10 woodworkers who I felt had been most influential in the field. The craftsmen I chose can be associated with the resurgent interest in American crafts that followed World War II, a period that saw an explosion of new styles and methods. All have contributed to the craft either through teaching or writing, or by introducing innovative designs or techniques. The older craftsmen are particularly important because they took the risks and built a market for their work when support for the independent designer-craftsman was nearly unknown. At this writing, all but Esherick are alive and creating.

During my research, I often went back and visited a craftsman several times to understand the mission behind his work. Each time I dropped in on Arthur Espenet Carpenter, he was trying something new with his work, exploring ways to incorporate various materials, colors and forms into his furniture. Wendell Castle's furniture has changed dramatically from when I first encountered his work in the mid-1970s. But as I considered it more carefully, I understood the logical progression of his style and themes. On the other hand, I saw little difference between a sample of bowls Bob Stocksdale turned

for his mother in the 1950s and his works of today, except that they have become thinner and lighter. Sam Maloof deliberates for months over a subtle change in a chair design he has been perfecting over 30 years, while Garry Knox Bennett needs to move on to a new project every few days.

I define woodworking in terms of a work's intended function as well as the material. The artisans discussed in this book accept the challenge of creating an attractive object that also fulfills a practical need. How they balance utility and aesthetics is a source of infinite and fascinating variation. Although they work predominately in wood, their approaches toward the material couldn't be more disparate. George Nakashima's reverence of wood, for example, springs from his Japanese heritage, which taught him to honor craftsmanship and respect nature. A more irreverent attitude is held by Garry Knox Bennett, who views wood as merely a convenient medium for his exotic designs and who disdains expressions of sentiment for the craft. Sam Maloof shuns metal hardware, while Wendell Castle freely mixes wood with precious metals, glass and other materials. James Krenov applies only finishes that preserve wood's innate beauty, while Arthur Espenet Carpenter paints his furniture in bright colors whenever the mood strikes him. Although the craftsmen are evenly divided between the two coasts, I found little evidence to support the stereotypes of the West's supposed preoccupation with form and the East's blind obsession with technique. All share a need to create rather

than simply construct.

Each craftsman has a different story to tell on how and why he became a woodworker. Both George Nakashima and Bob Stocksdale learned their craft while interned in American concentration camps during World War II— Nakashima for his ancestry (even though he was born in Seattle) and Stocksdale for resisting the draft (a rare stand, at that time, for a midwestern boy). Tage Frid's parents apprenticed him to a cabinetmaker because (so he claims) he wasn't smart enough to attend college; Frid eventually became full professor at the prestigious Rhode Island School of Design. Wharton Esherick and Sam Maloof began making furniture for their own homes, where it was seen by friends who persuaded them to sell their work. Marijuana was responsible for Garry Knox Bennett's transition from artist to craftsman. From sculpting roach clips (marijuana cigarette holders), he progressed to larger functional objects such as clocks and lamps and later to wood and metal furniture.

Despite their divergent backgrounds and styles, these men share in the struggle and satisfaction of being independent craftsmen. Rebelling against a fickle society obsessed with fads and assembly-line art, they retain their individuality and standards. They also maintain an unshakable faith in their own talents, even though it was often years before they could support themselves through their work. For them, woodworking is a means to personal as well as creative freedom.

Michael A. Stone

ACKNOWLEDGMENTS

This book would have been impossible without the cooperation of the nine living craftsmen whose careers I account. Each took time from his work to share his experiences and insight. I am grateful for their hospitality and friendship.

Nathan Rubinson, who supported this project from its inception, Donald L. McKinley, Lawrence and Alice Seiver, Miriam Phillips, Horace Hartshaw, Ed Ray and the late Nellie Lee Bok were among those who graciously provided information for the chapter on Wharton Esherick.

Among the many patrons who allowed me to invade their homes or who provided photographs of their collections, I am particularly grateful to York K. Fischer, David L. Davies, Robert Bruce Balbirnie, Charles and Louise Hallick, Horace P. Billings, Jay and Michelle Heldman, Robert Kaufman and the Rockefeller family. Mason Wells and Garretson W. Chinn of Garrett Wade Co. generously assisted with photography expenses. Lynn E. Farkas and George W. Stocksdale lent their collections of bowls to be photographed.

In conducting my research, I consulted numerous talented craftsmen around the country. I am especially grateful to David Ellsworth, Alphonse Mattia, Craig McArt, Morris Sheppard, William Keyser, Jr., Judy Kensley McKie, Lawrence B. Hunter, Jon Brooks, Ken Strickland and Ed Zucca.

Bob Hanson, Jonathan Pollock, G. William Holland, Donald L. McKinley and Esto Photographics Inc. took many of the fine pictures that illustrate the text. I am indebted to Alexander F. Milliken of Alexander F. Milliken Inc.; Rick and Ruth Snyderman of The Works and Snyderman Gallery; Warren and Bebe Johnson of Pritam & Eames; Jonathan Fairbanks of the Boston Museum of Fine Arts; Rick Mastelli of The Taunton Press; Paul Smith of the American Craft Museum; Dan Gordon; the National Museum of American Art of the Smithsonian Institution; and Judy Coady, Bernice Wollman and Warren Rubin of the Gallery at Workbench for providing photographs and assistance.

Special thanks to John Kelsey of The Taunton Press, one of the few true authorities on American woodworking, and Richard Kagan, a fine craftsman in his own right, for their advice and encouragement. My appreciation to Joanne Polster of the American Craft Council Library, Michael McTwigan and Steven J. Stone for their counsel. Thanks, too, to my editor, Laura Tringali of American Artisan Press, for all her help and to Clane Graves for his excellent art direction. Lastly, for her support and editorial criticism, my deepest appreciation and affection go to my wife, Louise Stone. It is to her this book is dedicated.

Sam Maloof's favorite design is his low-back dining chair.

INTRODUCTION

The dramatic pattern on the Fountain cabinet by Wendell Castle (1982) is sterling silver inlaid on an English walnut background. The hardware is also silver.

Each of the ten craftsmen discussed in *Contemporary American Woodworkers* has played an important role in the American Crafts Revival. Wharton Esherick, the oldest of the group, was born in 1887 and died in 1970. His career, which is linked back to the Arts and Crafts Movement of the late nineteenth and early twentieth centuries, inspired modern craftsmen in the early days of the Revival in the 1960s. George Nakashima, Bob Stocksdale, Tage Frid, Sam Maloof, Arthur Espenet Carpenter, and James Krenov—craftsmen born in the first two decades of the twentieth century—are now the elders of the Revival. All are revered, though not necessarily by each other. They started their careers when there were few independent woodworkers and it is largely through their efforts that the ranks of amateur and professional woodworkers have swelled. Wendell Castle, Garry Knox Bennett, and Jere Osgood—all born in the 1930s—are members of the first generation of woodworkers brought forth by the Revival itself and exercise a great deal of influence on the heirs to the tradition of studio woodworking.

Just as Arts and Crafts styles protested the mechanical mass production of Victorian ornament, so Crafts Revival styles of the 1960s and 1970s may be read as a high-minded reaction against the color and gloss of the then-fashionable Pop and High-Tech styles of decoration. But parallels between the two movements are not confined to aesthetic attitudes. The lives led by many Arts and Crafts workers were shaped by a desire to escape from the industrialized Victorian world, and many American craftsmen in the 1960s and early 1970s engaged in a similar search for "alternative lifestyles."

The notion that society is misusing its technical and natural resources by producing too many consumer goods is unlikely to commend itself to the poor; those who have had little opportunity to enjoy the products of the machine will not soon become disenchanted with them. It is therefore no surprise to find that many Crafts Revival workers are, like their Arts and Crafts predecessors, middle class in education and outlook. No surprise but still something of an anomaly, since craftsmen make things with their hands and historically this has placed them low in the social hierarchy. In antiquity, craftsmen were often slaves. In Europe, from the Middle Ages on, the guild system boosted the incomes of entrepreneurial masters at the expense of the journeymen who actually produced most of the goods. Only painters, sculptors, and architects managed, by the seventeenth century, to raise their crafts to the status of professions. The reason seems clear enough. These arts demanded invention rather than the repetition that is the lot of most craftsmen, and seventeenth-century inventiveness sprang from the literary and, in the case of architects, the mathematical education available only to gentlemen. Artists were learned; craftsmen, trained in the apprenticeship system, lacked both book learning and gentility.

This picture of the status of the craftsman little resembles how the nineteenth century English Arts and Crafts prophets—Pugin, Ruskin, and Morris—imagined things to have been. Looking at the flood of cheap and ugly consumer goods coming from the factories, they made the assumptions that craftsmanship was dying, that the handmade was necessarily superior to the machine made, and that what they considered to be the superior work of old-time craftsmen showed they had led happy, fulfilling, and usually rural lives in a society that recognized their worth. What could be more delightful, indeed more moral, than to emulate these paragons?

Those English followers of the prophets who moved to the simple life of the countryside to put their mentors' principles into practice were for the most part well-educated and well-heeled gentlemen. They had grown up outside the craft tradition, they had not apprenticed, and they often brought in anonymous urban craftsmen to help them out. The records of the best group of Arts and Crafts woodworkers, the Barnsley brothers and Ernest Gimson, show they made unrealistically low profits that were acceptable only because of the private incomes they enjoyed. In the United States, the most distinguished Arts and Crafts woodwork was that produced to the designs of the Greene brothers, who were architects to a wealthy clientele. The fittings and furniture were made with little regard to cost. By the 1920s, when Wharton Esherick took to woodworking and a rural way of life, the influence of the Arts and Crafts Movement had waned and craftsmanship in wood was dying.

Wharton Esherick used negative space and geometry in this oak and leather chair (1940).

Preceding page: home of Mira and Jonathan Yarnall, George Nakashima's daughter and son-in-law. Architecture and furnishings designed by Nakashima in 1970.

Even those topflight urban cabinet-making shops (so much despised by Ruskin and Morris), that survived into the twentieth century by producing reproductions of eighteenth-century furniture, were succumbing to the economic upheavals of the era. Of the various crafts that the Arts and Crafts Movement had fostered, only studio pottery, which turned to traditional Japanese modes for inspiration, retained its vigor between the 1920s and 1940s.

Japanese influence would also be felt in Crafts Revival woodworking, but traditional Scandinavian cabinetmaking would prove to be of prime importance. During the 1930s the Scandinavians had made Modern Movement furniture of wood, rather than the steel favored by the more highly industrialized nations. After World War II Scandinavian furniture remained functional in design but simply sculpted, somewhat biomorphic forms began to appear, and there was a revival of traditional exposed joinery. The furniture was made in small factories in which far more handwork was undertaken than would have been considered practical on the production lines of American plants. Scandinavian Modern furniture, which flowered through the late 1940s and 1950s, combined craftsmanship with function and in its simplicity and quality expressed the egalitarian ideals of the Scandinavian nations. For a moment something like William Morris's ideal, craft-based, socialistic society seemed to have been achieved and, as Morris would have wished, it was producing for a mass market goods that met most Arts and Crafts criteria.

That was something he had never been able to do.

All the elders of the Crafts Revival discussed in this book—George Nakashima, Bob Stocksdale, Tage Frid, Sam Maloof, Arthur Espenet Carpenter, and James Krenov—set up shop as independent craftsmen during the decade following World War II, and the work of each of them shows the influence of the Scandinavian Modern style. Nakashima, Stocksdale, Maloof, and Carpenter were pretty much self-taught, but Frid and Krenov, who received formal training as cabinet-makers, were taught in Scandinavia. In Nakashima's work the Scandinavian influence shows in the simple box forms of some of his casework, the simplified Windsor style of his chair backs, and the plainly turned, tapered legs of his stools and coffee tables. The other elders also work in personal versions of Scandinavian Modern, but within this unity of style is considerable diversity. Except for Nakashima, the elders are not designers by training or, to judge by some of their statements, inclination. However, their work has an improvisatory quality that can be appealing.

The lives of five of the six elders discussed in this book (the exception is Tage Frid) suggest a common independence of spirit. Few seem to have fought in the war, though Maloof was in the army, and all seem to have been at loose ends during the post-war period. Each turned to woodworking because it allowed him a freedom that was otherwise hard to come by. Nakashima and Krenov explicitly, perhaps others implicitly, sought a mystical fulfillment in working with wood. But otherwise they were a scattered band who did little to seek each other out. Unlike Arts and Crafts workers, they lacked spokesmen and leaders and the social connections that might have given them ready access to, and influence over, a moneyed clientele.

Tage Frid stands apart from the other elders in that he grew up in a craft tradition and learned cabinetmaking through the apprenticeship system. He was plucked from that life by an invitation to become America's first cabinetmaking academic. As a teacher and a teacher of teachers his influence has been enormous. He taught sound technique to the first generation of Crafts Revival woodworkers, and through his writings has reached far beyond the students who attended his classes. Frid's books and articles, which are eminently practical, stand in contrast to the writings of Nakashima and Krenov. Frid instructs while Nakashima philosophizes; Frid is the craftsman turned academic while Krenov is the intellectual turned craftsman. Among the elders there are profound differences in attitude in spite of the relative homogeneity of their woodworking styles. Nakashima, Krenov, and Carpenter have all defined their positions on matters of process and aesthetics; Stocksdale, Frid, and Maloof center their concerns on efficient production. The first group would have felt at home as members of the Arts and Crafts Exhibition Society; the second might have found it highfalutin'.

The history of the American Crafts Revival has not yet been written, but when it is we may expect it to show parallels between the burgeoning of studio crafts and the transition from the conformist 1950s to the hip 1960s. The 1960s brought the United States into an era in which large numbers of well-to-do young adults found it possible to view their parents' affluence as a sordid boon and to avoid the expected entry into business, the professions, or housewifery. The trio of Wendell Castle, Garry Knox Bennett, and Jere Osgood are just a bit too old to be counted as members of this group, but they are of a later generation than the elders and are products of the Revival rather than precursors of it.

Osgood and Bennett represent two extremes of Crafts Revival woodworking attitudes. Osgood, academically trained by Frid, himself became an academic. Entering the field early in the 1960s he developed, and has stayed with, his personal version of Scandinavian Modern. He represents Crafts Revival orthodoxy. By contrast, Bennett is a latecomer. Irreverent about wood and woodworkers, he barged into furnituremaking in the late 1970s at precisely the point when the Early Crafts Revival style (squared-off Scandinavian Modern-cum-Shaker with overexposed joinery) was becoming a bore. He is a heretic and an iconoclast. Wendell Castle lies between these extremes. He took over Frid's position at the School for American Craftsmen when the older man left to teach at the Rhode Island School of Design, and began a teaching career in which he is still engaged. But unlike Frid, Castle is primarily a designer and the pieces he designed in the 1960s and 1970s rejected the fundamental idea that furniture is constructed of sticks and boards.

A serious, independent woodworker setting up between, say 1960 and 1975 had stylistic and marketing choices to make. He could devote himself to supplying the apparently interminable demand for reproductions of eighteenth-century furniture; he could choose to work in the Early Crafts Revival style (probably the most usual choice); or he could try doing "something different," thereby feeling he was expressing his individuality and being able to offer his work as art rather than craft, which could make good economic sense. Wendell Castle, the most visible of this adventurous group, found that the key to novelty was in the rejection of the logic of rectilinear forms and bilateral symmetry that allowed easy, economical construction. For these he substituted the illogic of curvilinear, often biomorphic forms that owed something to Art Nouveau and something to the 1930s and 1940s work of such sculptors as Moore and Noguchi. Writing about this Biomorphic style, the roots of which, he implied, were in the techniques of laminating and carving rather than in historical precedents, Castle suggested that the woodworker would "look almost everywhere and anywhere for his basic ideas except in the field of existing furniture." Ironically, these words did not appear in print until 1980, immediately before Castle's historicizing adoption of an Art Deco furniture style.

More recently Wendell Castle has said that his Biomorphic style was right

A collage of shapes, colors, and textures, this elaborate desk with lamp (1984) illustrates Garry Knox Bennett's talent for balancing elements. The wood is California walnut and the lamp is copper-plated brass.

for the 1960s, that he could get by with it in the 1970s, but that it was wrong for the 1980s. These are the views of a man who has fallen in with fashion, a trait he shares with the great cabinet-makers of the past who made it their business to keep abreast of what was new. But Castle has done more than merely switch styles. Like other studio work of the 1960s and 1970s, Castle's furniture was a protest against fashion. His went further and challenged the comfortable assumption that, fashion aside, all furniture was made up of a limited set of fundamental forms. But in the 1980s Castle has joined himself to what is fashionable. His Art Deco furniture responded to an interest among decorators in reviving a 1920s style that had been swept aside by the Modern Movement and his more recent, architecturally inspired pieces reflect the widespread interest among architects in reviving classical forms. Castle is, as he always has been, very much with it, but somehow the "them" of the 1960s have become the "us" of the 1980s.

Broad social shifts are, of course, behind this. The 1980s is not an age of protest. The attractions of the simple life seem to be fading and it is a sign of the times that New York's ultra-trendy *Metropolis* magazine regularly runs color ads for Post-Modern furniture made by studio woodworkers. In announcing the withdrawal of support for its Program in Artisanry in January 1985, the President of Boston University noted the program's declining enrollment and added that "the program got started in the '70s when there

was a 'Back-to-Nature' movement and lots of people wanted to be self-sufficient artisans. The appeal of this has waned in the '80s. The rat race is more appealing to students now." The American Crafts Revival may have run its course.

If the movement is dying, the legacy it will leave to woodworking is great. The Revival brought a great many talented amateurs to woodworking and as leisure time increases over the next decades their number is unlikely to diminish. Among professionals we may expect to see a thinning of the ranks of the less talented. What will happen to the leading professionals is a question. There is probably never going to be much room in the market for meticulous, costly, one-of-a-kind furniture sold through art galleries. At another level, there may be a place for the designer/craftsman/entrepreneur who can design well (which may mean fashionably), manage a shopful of assistants, and be a skillful salesman and businessman operating his own urban showroom. Are there such Chippendales in our future?

A. U. Chastain-Chapman
Northfield, Massachusetts
July 1985

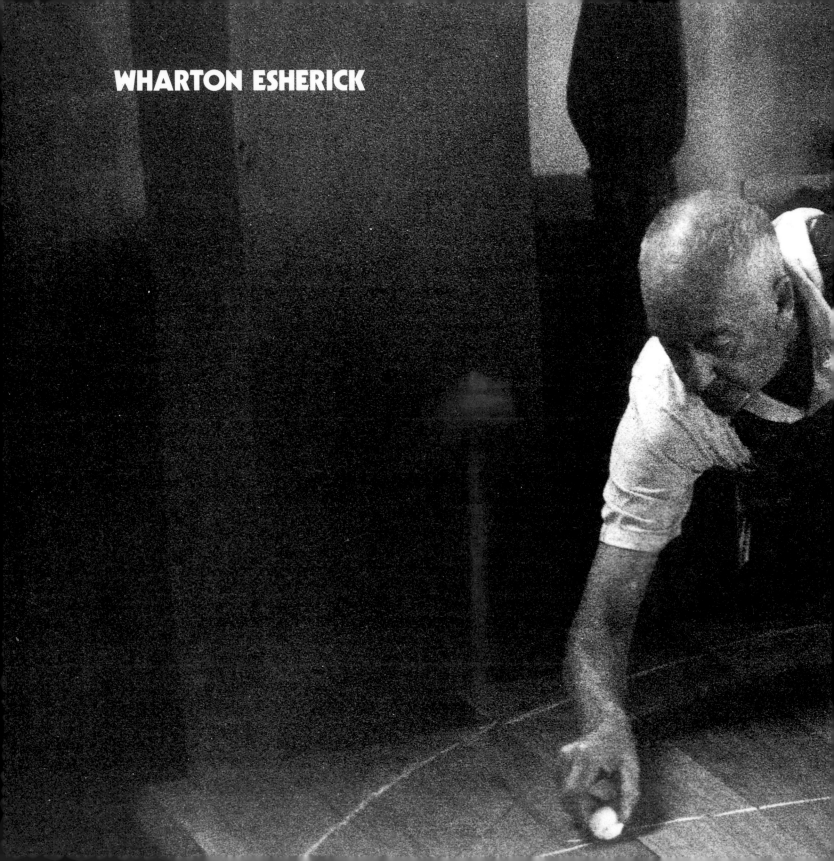

WHARTON ESHERICK

Wharton Esherick, artist turned craftsman, was among the first Americans to apply the aesthetics of contemporary art and design to the woodworking craft. When Esherick began to make furniture in the early 1920s, mass production was unchallenged and most pieces of furniture were nostalgic reproductions or European imports. Yet Esherick stubbornly adhered to the ideals of individuality and craftsmanship left over from the recently expired arts-and-crafts movement. Art nouveau, cubism and art deco were among the movements Esherick incorporated into his designs, thereby elevating furniture to a level of artistic expression then unknown in this country. Gradually Esherick developed his own style of fluid, organic forms inspired by the wood itself. This reliance on the material for ideas characterized his work throughout his long career. "Trees were the very life of Wharton," said Louis Kahn, the late Philadelphia architect who collaborated with Esherick on the construction of his unusual pentagonal workshop. "I never knew a man so involved with trees. He had a love affair with them; a sense of oneness with the very wood itself. . . ."

Today many young craftsmen unwittingly follow Esherick's lead, never knowing that Esherick pioneered the way decades before. But many of the country's leading designer-craftsmen recognize their debt. Wendell Castle, for example, acknowledges that he first came to realize that furniture could be art when he saw Esherick's work.

According to Castle, "Esherick taught me that the making of furniture could be a form of sculpture; Esherick caused me to come to appreciate inherent tree characteristics in the utilization of wood; and finally he demonstrated the importance of the entire sculptural environment." Arthur Espenet Carpenter, whose innovative designs and independent life-style have influenced many young craftsmen in San Francisco's Bay area, claims that Esherick's designs dared him to escape the limitations of traditional furniture forms: "It was in the mid-fifties when I first saw Esherick's work and began to see the possibilities of sculpting the edges of my furniture and eventually making it asymmetric." Sam Maloof, whose career began about 25 years after Esherick's, dubbed him "the dean of American craftsmen." He remembers the encouragement Esherick offered him when they first met in 1957, a time when furnituremakers who created their own designs were few and far between.

Esherick was born in 1887 to a well-to-do Philadelphia family. Against his parents' wishes, he attended manual-training high school and went on to study painting at the Philadelphia School of Industrial Art (now the Philadelphia College of Art) from 1907 to 1908 and the Pennsylvania Academy of the Fine Arts from 1909 to 1910. He became disenchanted with the program, however, and withdrew from the school just six weeks before graduation. "I could paint like Sargent painted, like old Vermeer. . . . But I couldn't paint like Wharton Esherick." He retained his mistrust of formal education—years

Esherick began building his studio in 1926, using sandstone and oak from the surrounding Pennsylvania countryside. He moved into the studio shortly after the main section, which housed the workshop and a bedroom in the loft, were completed in 1927 and lived there until his death in 1970. Throughout his life, he continued to expand and refine the structure. In 1941, he added the tall wooden section that contains a dining area and second bedroom. Fulfilling his dream for a silo, he erected a concrete tower in 1966 containing a kitchen and dressing room. The silo is covered with a fresco of pigmented cement: Esherick guided two masons in the application of each color to compose an enormous abstract painting inspired by the local landscape.

Esherick at age 78.

Esherick carved the handles and railings of his studio so they would feel as though they had been worn from years of use, as in this ebony door latch (1926).

The studio centers around a twisting red-oak staircase made in 1930, which connects the main floor with the dining area and Esherick's bedroom. Heavy, faceted steps are tenoned into the sculpted 9 ft. 9 in. center post and locked with 8-in. bolts.

later, when asked to teach, he responded: "I make, I don't teach. Bring your class out here. If they can't learn by looking at the thing, all my chattering will never teach them."

Esherick's first job was as an illustrator for two Philadelphia newspapers, but advances in photographic reproduction ended his career two years after it began. In 1912, he married Letty Nofer and moved to a farmhouse in Paoli, Pennsylvania, to pursue his painting full time. His first involvement with wood occurred in 1919 when he carved the frames of his neo-impressionistic paintings to make them more marketable. His friend, writer Sherwood Anderson, once told him his frames were better than his paintings.

During the 1920s and early 1930s, Esherick combined his love of drawing with his new interest in wood by carving woodcuts for printing. His prints illustrated magazines such as *Vanity Fair*, *The Century* and *The Forum* as well as nine books, including two volumes of Walt Whitman's poems, *Song of the Broad Axe* (1924) and *As I Watch'd the Ploughman Plowing* (1927).

Esherick soon broadened his use of wood to include sculpture, the genre he felt afforded him the greatest artistic freedom. Although he also experimented with stone and clay, wood over 15-foot-high Twin Twist carved in 1940 from a red oak log (p. 15). But while Esherick's sculpture was shown extensively in the Philadelphia area, it received only limited national recognition.

Like many postwar craftsmen, Esherick was entirely self-taught. His first furniture was for his own use, because "I didn't like the furniture we had at home." His early work emphasized surface decoration that was suggestive of his woodcuts—for example, the lower and center panels of a 6½-foot-tall drop-leaf desk he made in 1927 has low-relief carvings of the trees he saw from his windows. On the top doors, two turkey buzzards swoop over the treetops. The primitive, bulky form of the desk has more in common with the country furniture of Esherick's rural Pennsylvania than it does with the graceful designs of his later years. Another of Esherick's early experiments with furniture was decorating the sides of an antique walnut chest with stylized art-nouveau carvings of his house and property. Friends and patrons who visited Esherick's home urged him to make furniture to sell. By the mid-1930s, he had pragmatically abandoned painting and woodcuts in favor of furniture, sculpture, interiors and utilitarian objects—using these to express his artistic vision.

During the early 1930s, Esherick's preoccupation shifted from surfaces to form, and the furniture and interiors he created during those years incorporated the bold geometry of cubism and art deco. His finest and most ambitious commissions of that time were for the home of Judge Curtis and Nellie Lee Bok of Gulph Mills, Pennsylvania, which was called "one of America's outstanding domestic interiors," by *The Britannica Encyclopedia of American*

WHARTON ESHERICK

Bulky forms and relief carving characterize Esherick's early furniture, including this 6½-ft, drop-leaf desk (1927) of red oak used to store his drawings and woodcuts. The leaf opens on a wooden hinge held by a metal pin.

Art. Esherick designed and constructed the interiors of the music and living rooms and created fireplaces, archways, furniture and sculptures throughout the rest of the house. The living room contains space for 8,000 books, including a hidden shelf that pulls out of a bookcase to hold the Judge's large law books. An oak portal composed of a concave arrangement of concentric geometric shapes connects the book and music rooms. The motif is repeated around the fireplace, except there the pattern is convex. Similarly, Esherick shaped the handles on the doors between the rooms in reverse images of each other.

Among the furnishings he created for the book room was an arching library ladder to commemorate the Judge's unsuccessful campaign for district attorney. The Judge, who ran as a democrat, had switched political parties shortly before the election, and Esherick satirized his dual allegiance by topping the ladder with caricatures of an elephant and donkey. (The staunchly Republican Esherick told the Judge he could climb up and down its steps to symbolize his vacillating political loyalty.) For the building's exterior, Esherick designed a stone staircase to wind around the twisting chimney.

The masterpiece of the house is a spiral staircase made up of interacting arcs and angles. Solid white-pine beams retrieved from a bridge believed to date back to the American Revolution form the steps. The first 10 steps stand free; the eleventh attaches to the rear wall. To stabilize the spiral, a curved iron band passes through the inner edge of

Using contrasting convex and concave patterns in an art-deco motif, Esherick built a fireplace mantle and door arch in white oak for the book room of the Bok house. The hearth is stone covered with studded copper. The music-room fireplace in the background is plastered stone and the walls and ceiling are paneled with white-pine boards joined along their lengths with a tongue and groove.

each step. Esherick painted the back wall black to accentuate the rich, golden color of the seasoned pine and placed a light behind the stairs to illuminate the pattern of radiating lines and shadows formed by the underside of the steps. Similarly, light emanates from hidden crevices within sculptured wall panels in the music room to establish a mood and create an illusion of depth.

In developing the design of the Bok staircase, Esherick created a 17-inch scale model. He made occasional use of models for large commissions to show designs to clients and to test ideas.

Another notable piece commissioned during the 1930s, a lady's corner writing desk, stands as one of the finest examples of cubism ever applied to furniture. Its multiple surfaces resemble the facets of a gemstone. Three triangular panels fold back to reveal a complex interior of drawers and compartments in the same style of intersecting planes. Like the Bok home, the desk illustrates Esherick's effective use of light and shadow. The acute angles between the drawers formed by the handholds—Esherick detested extraneous knobs—define and separate the drawer fronts

Occasionally, Esherick constructed models for large commissions, including this white-pine miniature (1935) of the Bok staircase

Esherick topped this library ladder (1935) with caricatures of an elephant and donkey to commemorate Judge Bok's bid for public office. The ladder also doubled as a tea server.

and enhance the piece's three-dimensionality (see pp. 12 and 13).

Esherick's reputation grew after he collaborated with architect George Howe on one of the 16 interiors in the "America at Home" exhibition for the 1939-40 New York World's Fair. Entitled "Pennsylvania Hill House," the display included an asymmetric table with a black phenol top, chairs, sofa, wall paneling and the spiral staircase transplanted from Esherick's own studio. Continuing the use of geometry in his designs but with a new openness, he constructed the table and chairs from thin sticks of hickory that frame blocks of space. The sofa also served as a room divider. Like the more recent end table/chest, it contains curved drawers exemplifying his economic use of space and Esherick's ability to design pieces that satisfy multiple functions.

Solid pine beams from an old bridge once spanning the Delaware River form the steps of this magnificent spiral staircase commissioned in 1935 by Judge Curtis and Nellie Lee Bok. Adjacent steps are joined with wedged dowels and an iron band passes through the edge of each step to maintain the curve. The soft pine had to be sanded repeatedly over the years, especially during the 1940s when women's spike heels were in fashion, until the current owners were forced to carpet it. Esherick later admitted oak would have been a better choice. To the right of the stairs stands Esherick's sculpture The Actress (1939), carved from the crotchwood of a cherry log he rescued from his fireplace. The figure was inspired by a photograph of Esherick's daughter preparing for a play at the Hedgerow Theatre.

During the 1940s, Esherick settled into a style of freeform shapes and clean, sensuous lines that he practiced for the rest of his career. "Some of my sculpture went into the making of furniture. I was impatient with the contemporary furniture being made—straight lines, sharp edges and right angles— and I conceived free angles and free forms; making the edges of my tables flow so that they would be attractive to feel or caress."

In later work, he relied upon the material for design ideas rather than on prevailing artistic trends. He sketched directly on the rough boards to take full advantage of their innate strength, figure and color. "The wood has an influence on you. You can't fool with wood," he said. "I begin to shape as I go along. The piece just grows beneath my hands. I treat furniture as though it were a piece of sculpture. I dig up what I do out of my soul."

Esherick's spiral library ladder demonstrates the use of natural forms found in his later work. Steps sprout from the center post like leaves on a stem. Wood cut from around the joints softens the transitions and makes the parts seem to grow together. Other examples are the dictionary stand and music stand designed during the 1950s. The sweeping uprights of the dictionary stand are formed by curved tree branches, obviating the need to bend the wood.

Esherick's working philosophy is perhaps most visible in his studio. He described his goal of creating a totally integrated and sculptured living space as "the establishment of a mood congenial

Eventually, Esherick allowed the wood to dictate the form of his furniture. The steps of this cherry and hickory library ladder (1969) are tenoned into the center post and dovetailed to the legs. The curved members between the legs are secured by rabbet joints and wooden nails.

WHARTON ESHERICK

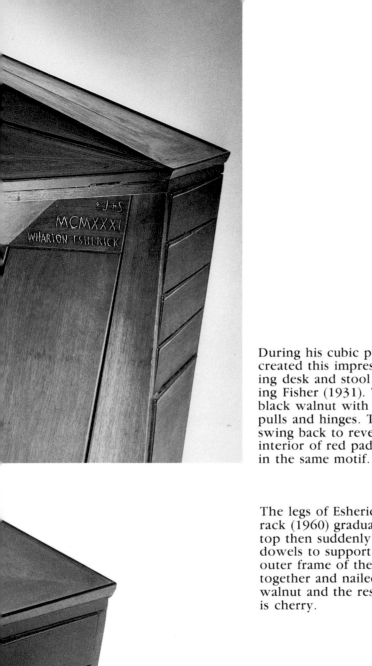

During his cubic phase, Esherick created this impressive corner writing desk and stool for Helen Koerting Fisher (1931). The exterior is black walnut with ebony bead, pulls and hinges. Three panels swing back to reveal an intricate interior of red padauk and leather in the same motif.

The legs of Esherick's famous music rack (1960) gradually widen at the top then suddenly taper into long dowels to support the rack. The outer frame of the rack is tenoned together and nailed. The shelf is walnut and the rest of the stand is cherry.

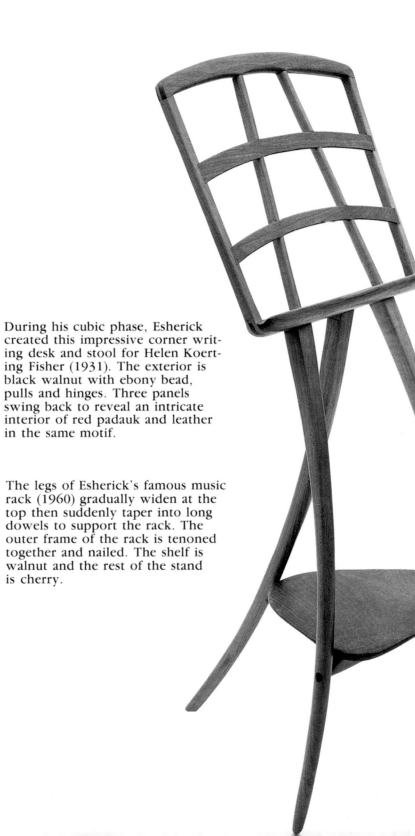

WHARTON ESHERICK

Esherick created this asymmetric hickory table and chairs (1939) for one of 16 interiors in the "America at Home" exhibition of the 1940 World's Fair. The wood is hickory; the table has a black phenol fiber top and each of the chairs has a leather seat.

Illustrating his imaginative use o found objects, Esherick construc 48 chairs in 1930 composed of hickory hammer handles joined with round tenons. He angled th back legs so they wouldn't tip c no matter how far back the sitte leaned. The seat is woven from canvas machine belting and painted.

with the existing or created environment," a sentiment shared by architect Frank Lloyd Wright. Using native sandstone and oak, Esherick began construction in 1926 just up the hill from his farmhouse, on a ridge then known as Mount Misery. Every feature of the studio, from the gentle taper of the stone walls as they rise from the foundation and the subtle sag in the roof to the tiniest accessories like carved light pulls and switch plates, was lovingly shaped by hand. Over a period of 40 years, Esherick carefully molded every detail—floors and walls, furnishings and doors, staircases and railings, lamps and utensils, even the toilet seat and coat hooks, into a unified environment. The studio centers around a massive red-oak spiral staircase that ascends from the main level like a branching tree. Bulky, faceted steps cantilever from a twisting

center post, giving the stairs a primitive and unpretentious quality (p. 6).

When the studio was completed Esherick left his family and made it his permanent home. As a man who needed to be alone with his work, he found it difficult to balance family and career. He worked seven days a week and rarely traveled. Miriam Phillips, his long-time companion, said, "Wharton would not even keep a pet because it would distract him from his work."

In 1972, two years after Esherick's death, his three children and many friends opened the studio to the public as a nonprofit museum. The Wharton Esherick Museum contains several hundred examples of work spanning his 60-year career. Included in the collection are models, prototypes and designs he repeated in series, such as his widely imitated three-legged stool. Esherick and his assistants formed the seats from interesting scraps and wildly figured pieces of crotchwood that they found on walks through the woods, and although they created several hundred stools, as in all Esherick's work, each one is unique.

Illustrations of Esherick's ingenious use of found objects can be seen throughout the museum. Once he bought a barrel of tool handles at an auction. When the impoverished actors of Hedgerow Theatre, a local repertory company for whom he designed stage sets, asked him to make some chairs, he constructed them from these handles. By teaching the actors how to finish the frames and weave the seats from canvas machine-belting, the cost was minimal. On another occasion, he fashioned a set

Esherick was years ahead of his time with this organic sculpture called Twin Twist (1940). It was carved from a red-oak log and towers over 15 ft. high.

of chairs from wagon wheels for a harness room in Mount Kisco, New York. In his own home, he turned a twisted mastodon tusk into a staircase railing.

His techniques were simple and honest. He found joinery tedious and delegated it to his shop assistants Jon Schmidt, Bill McIntyre and Horace Hartshaw—all trained cabinetmakers. Hartshaw remembers Esherick as a benevolent employer who never raised his voice, preferring instead to express his displeasure through humor. Esherick relied on his assistants to supplement his limited knowledge of technique while he concentrated on designing and shaping. Less than a year before his

Esherick used the pear shape of this desk top in many of his desks and tables. Deeply concerned with how his furniture felt when handled, he rounded the edges to give them a sensuous quality. This desk (1958) is made of walnut and cherry.

death at the age of 82, he told Sam Maloof, "I'm still shaping the seats of the stools. The boys just don't get the hang of it."

Esherick preferred to work in native woods—cherry, walnut, oak, dogwood and hickory, among others—because he best understood their properties. "If I can't make something beautiful out of what's in my backyard," he often said, "I'd better not make anything."

His long relationship with his wood supplier, Ed Ray, was a craftsman's dream. Ray, who also furnished George Nakashima at one point, called Esherick whenever he opened an especially attractive log. Esherick related one incident: "When he cut into an oak tree and found it was curly oak, he said to himself, 'I'm not giving that to some of these fellows to make steps and stuff like that—this goes up to Esherick.'"

Esherick used panel construction to create the shapes of his curved sofas

and cabinets. The panels were screwed to a batten that formed the curve, but were left unglued to permit expansion and contraction. The boards were connected by plywood splines that fit into slots along the edges. He employed a similar technique for wall paneling. In the living room of Lawrence and Alice Seiver of Villanova, Pennsylvania, Esherick used long slats of poplar like broad strokes of a painter's brush to create a window screen and paneling over the fireplace.

Esherick disdained romantic notions about the craft and used machinery wherever possible. Most of his shapes came from two bandsaws—one fabricated by Schmidt from two bicycle wheels. His shop also contained a stationary belt sander, jointer, thickness planer, drill press, table and radial-arm saws and a collection of gouges and rasps for final shaping. "This business of hand-done is a lot of nonsense," he once declared. "When they talk of made-by-hand, it's silly. It's made by hand, of course. But more, it's made by heart and head."

Esherick's work has been exhibited throughout the world and is found in the permanent collections of the Metropolitan Museum of Art, Pennsylvania Academy of the Fine Arts, Whitney Museum of American Art, Philadelphia Museum of Art, American Craft Museum and many others. The Architectural League of New York presented him with their Gold Medal of Honor in Design and Craftsmanship in 1954 and the American Institute of Architects posthumously awarded him their Craftsmanship Medal in 1971.

Esherick used broad slats of poplar like brushstrokes to create wall paneling and a window screen in the 1960s for the home of Lawrence and Alice Seiver. The boards are joined with painted plywood splines that fit into a groove routed along the edges of each plank. This permits the wood to expand and contract with changing humidity. He used a similar technique to construct the curved cherry sofa (1958), but glued one side of the spline in place so it wouldn't slide out.

One of Esherick's last pieces was this walnut chest/table (1969), which toured the country in the "Objects USA" exhibition.

GEORGE NAKASHIMA

More than anyone else, George Nakashima embodies a movement in contemporary woodworking in which the craftsman allows the wood to dictate the form and function of the object. To Nakashima, wood is sublime and design is relegated to the mostly technical process of engineering. Nakashima's genius is his ability to determine the most appropriate use for the magnificent lumber he acquires from around the world. "Each flitch, each board, each plank can have only one ideal use," he says. "The woodworker, applying a thousand skills, must find that ideal use and then shape the wood to realize its true potential." Nakashima harbors no interest in self-expression, preferring to "create beauty not art," a goal he sees lacking in most American crafts. His self-appointed mission is to preserve the splendor held within great hardwood trees by using them to create practical objects.

Nakashima's relationship with trees can only be described as spiritual; to him, trees personify a divine spirit that transcends civilizations, physically as well as metaphysically. Working with the bodies of these noble "beings," as he calls them, Nakashima links his life with nature. A disciple of the Hindu leader Sri Aurobindo, Nakashima chose woodworking as his *Karma Yogin*—his "yoga of action." He describes the relationship between his personal philosophy and profession in his book, *The Soul of a Tree:* "We work with boards from the trees, to fulfill their yearning for a second life, to release their richness and beauty. From these planks we

Nakashima standing among the great timbers in one of his woodsheds. This photograph was taken in 1962.

fashion objects useful to man, and if nature wills, things of beauty. In any case, these objects harmonize the rhythms of nature to fulfill the tree's destiny and ours."

John Kelsey, former editor of *Fine Woodworking* magazine, calls Nakashima's brand of craftsmen "Druids" for their religious idolization of the material. But their approach runs contrary to most contemporary trends in the craft. William Keyser, Jr., a woodworker who employs sophisticated bending techniques to create geometric designs, characterizes Nakashima's philosophy as "searching for God in a knothole." Other craftsmen such as Wendell Castle and Garry Knox Bennett

Side table with burl top.

In 1974, Nakashima's workshop furnished the entire interior of Governor Nelson and Happy Rockefeller's home in Pocantico Hills, New York. The dining table is made of two matching 13-ft. planks of East Indian laurel. The Conoid dining chairs are walnut.

represent the antithesis of Nakashima because they use furniture as a means of artistic expression—they appreciate wood, but mostly it's a convenient medium for their designs. They differ from Nakashima by the degree in which they impose themselves on the material. To Nakashima, furniture is most successful when the design is unobtrusive and the craftsman's ego does not intrude upon the natural beauty of the wood. Sculptural wood furniture he dismisses as "better suited to plastic."

Nakashima's philosophy toward his craft evolved from years of study and travel throughout the world, and extensive exposure to both Western and Eastern societies. The result is a blend of two distinct traditions: Japanese respect

Among his earliest furniture designs, Nakashima has produced this walnut armchair since 1948.

for nature and purity of craftsmanship, and American individuality and self-reliance. A man of immense energy and intellect, Nakashima has achieved an inner satisfaction that comes from knowing one's ability and destiny. He climbs the woodpiles like a man half his age, kept young by his passion for work. He is happiest in the woodsheds, surrounded by the great timbers he admires.

Nakashima was born in Spokane, Washington, in 1905. He spent much of his boyhood exploring the forests and mountains of the Pacific Northwest, especially the Cascade and Olympic mountain ranges, where he developed an ardent appreciation for the wilderness. His parents held prestigious positions in Japan before immigrating to America: His mother served in the court of Emperor Meiji and his father was of samurai lineage. They endowed

him with a deep pride in his ancestry and the value of family unity.

After attending local schools, at the age of 20 Nakashima made his first trip to Japan to visit relatives. Shortly thereafter, he studied architecture at the University of Washington and spent a year in France at the École Américaine des Beaux Arts in Fontainebleau on a scholarship before graduating in 1929. He went on to postgraduate training at the Massachusetts Institute of Technology and earned a Master's degree in architecture in 1930.

Following two years of work on various architectural projects in New York, he decided to see the world. The trip would last seven years. Purchasing a second-class ticket granting him worldwide passage for two years on a British steamship line, he departed for England and France. He held odd jobs for a year in Paris before traveling to Tokyo, with

GEORGE NAKASHIMA

The turnings and walnut slab seat of Nakashima's New chair (1950) are reminiscent of the Windsor chair of early America. Nakashima offers the chair with or without arms and rockers.

Simple in its design and classic in its proportions, Nakashima introduced the Frenchman's Cove table in the mid-1950s. This table is made of East Indian laurel.

brief stops in the Middle and Far East. In Japan, he found employment with the architectural firm of Antonin Raymond. Raymond was Frank Lloyd Wright's architect-in-chief for Tokyo's grand Imperial Hotel before he established his own company in Japan.

In 1937, Nakashima volunteered to design and construct a dormitory for the ashram, or spiritual community, of the followers of Sri Aurobindo in Pondicherry, India. This was his first experience with furniture design and production, although on a limited scale. After about a year, he resigned his position with Raymond to join the commune and become one of Aurobindo's many disciples, donating his services to the ashram. However, believing his fate was to rejoin modern society to practice the principles he had learned in India, Nakashima returned to Japan in 1939 to resume his architecture career. At this time, he became engaged to Marion Okajima, an American English teacher living in Tokyo. They moved to Seattle in 1940 and married soon after.

Upon returning to the States, Nakashima set out with a friend to discover what modern Western architecture had to offer. They traveled down the West Coast surveying the work of Dailey, Wright, Harris and Neutra. The experience totally disillusioned him and sent him searching for a new career. "I got into furniture because of my negative reaction to architecture. . . . This stuff that was being done seemed exciting, but actually what really turned me against architecture was to have seen a Frank Lloyd Wright house under construction. I mean, just the hypocrisy

GEORGE NAKASHIMA

and the shabby construction . . . everything for a facade and a shape." Seeking a process he could control from start to finish, Nakashima chose furnituremaking as his career. In 1941, a local priest took an interest in Nakashima's work and allowed him to open a small business creating his own furniture in the basement of one of the church's buildings. Twenty-four years later, the priest was transferred to Katsura, Japan, where he wanted to build a church—

Nakashima's Connoid chair with cushion.

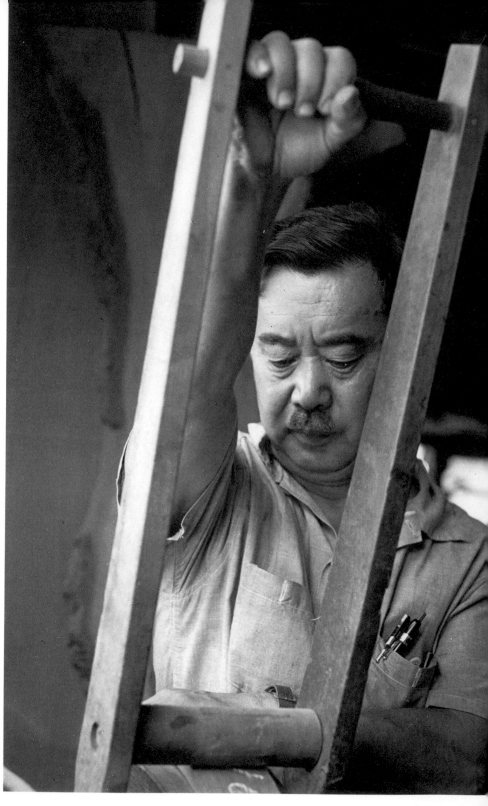

Nakashima donated his services and designed the structure and furnishings. Nakashima's daughter, also a trained architect, assisted in the design and supervised the construction.

The internment of Japanese-Americans during World War II sent Nakashima and his wife to the Minidoka Relocation Center at Hunt, Idaho. Surprisingly, his only bitterness toward his imprisonment concerns the hardships endured by his wife and daughter, who was born just six weeks before their confinement. While in the camp, Nakashima met a Japanese-trained carpenter about his own age who taught him to use traditional Japanese hand tools. "Using just scraps of wood left from the lumber scrap heap . . .," he says, "he and I got together . . . I was the designer and he was the cabinet-maker; and I was both his superior and his apprentice."

A job offer from Raymond, who had returned to the United States, helped free the Nakashimas in 1943 and brought them to rural New Hope, Pennsylvania. The town, located just north of Philadelphia in Bucks County, possesses a rich history of cabinetmaking dating back to Colonial times. It is also situated just 30 miles from Wharton Esherick's studio. But although the two craftsmen knew of each other, their rigid devotion to divergent styles and philosophies prevented them from pur-

The intricate top of this coffee table comes from the root of a giant redwood tree.

Nakashima in the early 1950s bending the back of a chair.

suing a relationship. In fact, Nakashima disassociates himself from the mainstream of the American craft movement, believing it to be preoccupied with fads and nearly devoid of good design and sincere craftsmanship.

Over the years, Nakashima's business blossomed from a solo operation in an open shed to a major enterprise employing about a dozen people working out of several buildings spread over 10 wooded acres. He initially made all the furniture himself, but the orders soon became more than he could handle alone. So he hired assistants to assume the cabinetmaking processes and devoted himself to design and customer relations as well as to the crucial tasks of selecting the logs, supervising their cutting and determining their use. Nakashima hasn't personally constructed an entire piece since. He permits his assistants the freedom to work at their own pace, for he believes "each woodworker is an individual craftsman, free to work out his own *sadhana,* spiritual training. . . . Each person can do what he finds most suitable within certain guidelines." Some of his assistants have traditional European training while others are local men trained by Nakashima himself. Three of his men have been with him since the 1950s; all of them share an intense pride in the quality of their work.

A cantilever table of English walnut and Conoid chair with Persianwalnut seat in Nakashima's house.

Nakashima's workshop produces about 75 stock designs that he has created and adapted over the years, including tables and chairs, benches and beds, cabinets and wall cases, chests and desks and lamps. Elements of both Early American and Japanese styles, the two major traditions in his background, are evident in his work. The lines in his designs are simple and precise, and Nakashima's masterful sense of proportion gives his work a classic look. According to Nakashima, "Furniture, though heavy, need not look bulky if proper proportions are used. . . . In a traditional Japanese house, the proportions used have been perfected over many generations. A variation of as little as half an inch in the horizontal can mean a devastating failure. In the same way, one mismatched board in a cabinet can be a disaster."

Nakashima often uses freeform slabs of wood for the tops of his pieces, especially of the tables. The bases are uncomplicated, practical structures that support and display the timber, which might be a fragile slice of burl or a convoluted root section. His chests and cabinets tend to be more traditional, although they may also have tops with natural, free edges. The Odakyu cabinet, with its delicate *asa-no-ha* (hemp leaf) grillework (p. 30), is an example of a traditional Japanese design that Nakashima incorporated into his work.

Nakashima's Conoid chair and New chair are among his best known designs. The Conoid chair sprung from Nakashima's desire to create a chair with only two legs; the supporting runners work well on carpeted floors.

Reminiscent of the Windsor chair of the eighteenth century, the New chair illustrates an Early American influence. The spindles are secured into the seat and top with round tenons and, like Colonial craftsmen, Nakashima shrinks the tenons with heat before insertion, which allows them to expand and lock in place. His shop produces about seven chairs a week. The construction process includes several concessions to

The size, shape and location of the butterfly joints are crucial in controlling cracks when joining long boards. By selecting a wood for the butterfly that contrasts with the top, the butterflies also become design elements. The butterfly is inlaid slightly more than halfway into the board and held with glue and screws.

This walnut cabinet (1948) rests on what Nakashima calls the "R" bench

This is a smaller version of the Frenchman's Cove table, originally commissioned during the 1950s to furnish a resort hotel in Jamaica.

production—the spindles, for instance, are turned on an automatic lathe by an outside supplier and jigs are used to cut the more complicated joints. All parts, however, are chiseled by hand to a clean, square and perfect fit.

The wood for these pieces comes from the great forests of the world—Nakashima uses oak burls from England, English and Persian walnut, Carpathian elm burl, French olive ash, root sections from the massive redwoods of California, and exotic East Indian laurel and rosewood. Yet the majority of the wood comes from the United States, mostly walnut and cherry. Frequently, the logs contain knots and imperfections, making them worthless by commercial standards. But this is the wood Nakashima treasures most, because it has the richest grain and most extravagant figure.

Nakashima ships most of his wood directly to a sawmill where he personally directs each pass of the giant bandsaw. "The sawing of logs is of prime importance," explains Nakashima. "Each cut requires judgment and decisions on what the log should become. As in cutting a diamond, the judgments must be precise and exact concerning thickness and direction of cut, especially through 'figures,' the complicated designs resulting from the tree's grain. If a figure is cut properly, the beauty locked in the tree will gradually emerge. If cut improperly, most is lost." Unlike most commercial lumber, which is grade-sawn or peeled off the log in thin sheets for veneers, Nakashima slices his logs straight through to take full advantage of the

Nakashima's chair shop produces five to seven Conoid chairs (mostly walnut) a week. Workers use heat lamps to shrink the spindles and fit them into sockets in the seat and rail. As the spindles expand, they lock into place—similar to a technique used by Colonial craftsmen.

Japanese styling is strongly evident in Nakashima's walnut Odakyu cabinet, named for the Japanese store where it was first exhibited in 1968. The intricate *asa-no-ha,* or hemp leaf, pattern is a traditional element in Japanese design—Nakashima also uses it in sideboards, lamps and cupboards.

log's width and graining.

After drying outdoors for one to two years, the wood goes to a kiln to extract further moisture. Then the boards are hauled to one of four lumbersheds on Nakashima's property. Finally, a few weeks before the wood is actually worked, it is brought into the workshop to acclimate to indoor temperature and humidity. Throughout this process, planks from the same tree are stored together in the same relationship to each other as existed in the log. This makes it easier to match grain patterns when several boards are needed to create large widths, as for tabletops.

Nakashima maintains a five- to 10-year supply of wood and constantly acquires new logs to replenish his stock—he stores only a third of his inventory on his property.

In 1984, one of Nakashima's loggers found a walnut log in Long Island, New York, which Nakashima says was the finest he had ever encountered. Because of its enormous size, five to seven feet in diameter, he hired lumbermen from California to saw it by hand. Inspired by its unusual beauty, he searched for a worthy purpose and decided to fashion it into altars for peace. He intends to donate these shrines to Hiroshima, and

An adaptation of the New chair, this walnut rocker (1963) has a free-form arm on one side.

Nakashima's own personal desk of English walnut (1959) is seen here in a corner of the Conoid studio.

possibly to an American city.

Like Nakashima's furniture, the workshops, showrooms, storage areas and living quarters on his property blend contemporary and traditional Japanese aesthetics. Nakashima planned and constructed all the buildings himself. The use of local fieldstone and timber integrates the structure into the rocky Pennsylvania countryside. Three buildings have parabolic roofs composed of thin shells twisted into sweeping geometric shapes. The roof of the Conoid Studio begins as a 40-foot arch on one end, then flattens to a straight line and finally contorts into a series of reverse curves at the back. This studio is Nakashima's showroom and office and contains examples of his work. The designs he created in this building are also titled "Conoid."

Nakashima built the Minguren Museum in the 1960s to display specimens of the world's extraordinary timbers as well as prototypes of his furniture and his personal collection of artwork and artifacts. The roof is a hyperbolic paraboloid. A mosaic by the late artist Ben Shahn, who was Nakashima's close friend, covers one of the outer walls.

Nakashima's home and a small building he calls his mountain retreat stand apart from the other structures. There are also separate shops for making tables, chairs and cabinets, and for finishing. Like most contemporary craftsmen, Nakashima freely uses machinery and power tools. His main shop contains tablesaws and bandsaws, thickness planers and jointers, stroke sanders and a lathe. He describes the balance he strikes between hand and machine: "As much as man controls the end product, there is no disadvantage in the use of modern machinery and there is no need for embarrassment. Gandhi and his spinning wheel were more quix-

Nakashima designed this walnut Mira chair in 1952 as a high chair when his daughter was a baby. The chair comes in a variety of heights and with or without a footrest.

otic than realistic. A power plane can do in a few minutes what might require a day or more by hand. In a creative craft, it becomes a question of responsibility, whether it is man or the machine that controls the work's progress." Numerous Japanese and Western hand tools line the walls. The shop has an atmosphere of relaxed efficiency. Because orders are scheduled one to two years ahead, it operates at a steady pace.

Reflecting his belief in close family ties, Nakashima's entire family participates in the business: His wife handles the bookkeeping and his son, Kevin, helps manage the operation. His daughter, Mira, a graduate of Harvard University and Waseda University in Tokyo, works closely with Nakashima in design and wood selection, and will probably inherit his mantle. Nakashima's children and four grandchildren live nearby.

His business and marketing techniques illustrate Nakashima's independence—he enjoys pointing out that he built his entire business without loans or mortgages. And, preferring to bypass galleries and other middlemen, orders come directly to the shop, either by mail or in person. He conducts open house for three-and-a-half hours on Saturdays to meet with clients and discuss their special requirements, often drawing them a rough sketch in the process. On occasion, customers select their own pieces of wood and Nakashima scribbles their names on the rough planks in chalk.

But Nakashima's achievements extend well beyond his furniture and his thriving business. He has done a number of architectural commissions: In addition to his work in India, he has designed religious buildings throughout the world, including the Church of Christ the King in Katsura, Japan (1965); the Monastery of Christ in the Desert, Abiquiu, New Mexico (involving several buildings completed in the late 1960s and early 1970s); and La Soledad, San Miguel de Allende, Guanajuato, Mexico (1975). He has also executed numerous commissions for interiors and furnishings for churches, synagogues, colleges and private residences. In addition, Nakashima has served as an industrial designer. Knoll International and Whitticomb-Mueller produced his designs during the 1940s and 1950s, and a small operation in Japan manufactures furniture under his guidance.

Nakashima's furniture has appeared

Conoid bench, coffee table and chair.

in exhibitions throughout this country and Japan, and is included in the permanent collections of the Boston Museum of Fine Arts, the Philadelphia Museum of Art, the American Craft Museum, the Saint Louis Art Museum and London's Victoria & Albert Museum.

His numerous awards include the Gold Medal for Craftsmanship from the American Institute of Architects in 1952; the Silver Medal of Honor in Design and Craftsmanship from The Architectural League of New York in 1960; the Honor Award for Inspired Creativity and Outstanding Sensitivity in Design from The Pennsylvania Society of Architects of the American Institute of Architecture in 1980, among others. In 1979 he was elected a Fellow of the American Craft Council and in 1983 the Japanese government honored him with the Third Class of the Order of the Sacred Treasure, an exceptionally high distinction.

Many have tried to imitate Nakashima without success, for they lack his essential rapport with the wood. To him, woodworking is more than a profession—it is "a way of life and a development outward from an inner core: something of the same process that nature uses in the creation of a tree."

BOB STOCKSDALE

To many American woodworkers, turner Bob Stocksdale epitomizes the traditional craftsman. His career of over 40 years has been devoted to producing the finest objects he can create—mostly bowls and trays. Undaunted by the craft's current focus on innovation and self-expression, Stocksdale strives for perfection in form and excellence in technique. His skill with the wood lathe has earned him universal recognition as a master of his craft. Having single-handedly revitalized the art of turning decorative bowls in this country, he has achieved legendary status among many professional and amateur American turners working today.

Stocksdale advanced wood-lathe artisanry in two significant ways: He increased and refined the possible shapes of turned bowls and he expanded the variety of woods used to make them. David Ellsworth, known for his artistic hollow turnings, believes Stocksdale should be designated a "national living treasure." Says Ellsworth, "I think of Stocksdale, the turner, as I do Edward G. Robinson, the actor: consistent excellence and, most of all, he is *The* Man."

Like bowls from a potter's wheel, Stocksdale's forms exude the classic grace and precision of fine ceramics. While attending an exhibition of early Japanese pottery at San Francisco's Asian Art Museum, he noted the resemblance to his own shapes and wryly

Stocksdale retained the natural edge of this bowl of California black walnut burl (1980).

remarked, "these people are wonderful at copying my designs!" Although he has experimented with more elaborate shapes, Stocksdale is most successful when he focuses on simple spheres and curved silhouettes. Yet within these seemingly limited forms, there is room for infinite variation. The serpentine or S-shaped profile, for example, with its double-reverse curves, has become Stocksdale's trademark and appears in many of his decorative bowls. The "bird's mouth" shape, so called because it resembles the beak of a baby bird with its head thrown back and its mouth wide open, is another example. The dip in the bowl's rim comes from the natural curvature of the log—the grain runs across the bowl instead of up and down. Sometimes, when using a dark wood such as walnut or muninga, it's possible to leave a contrasting ring of lighter sapwood around the rim. Although the dip creates the illusion that the bowls are oval, they are as round as Stocksdale's visually round pieces.

Another way that Stocksdale varies his designs is with texture. He'll occasionally retain the natural edge of the wood to give the rim a freeform shape, or leave a ring of coarse striations, left by the bandsaw, along the center of the bowl. By adding a pedestal, the shape and stability of the piece can be further altered. But regardless of its design, diameter or depth, each piece possesses the fluency of form that affirms his instinct for the material and his singleness of purpose.

Rather than approach each bowl with a preconceived shape, Stocksdale

The bowl is walnut, which Stocksdale considers one of the easiest hardwoods to turn, and the mortar and pestle (1965) are of lignum vitae, a dense wood with distinctive light and dark grain patterns of sapwood and heartwood.

evolves the form as he turns. "I don't know where my shapes come from," he says, "they just grow as I turn." Every piece is turned freehand, that is, without the aid of jigs or patterns. The forms materialize almost magically as he glides the gouge back and forth along the spinning wood—mounds of long, neat shavings accumulate at his feet as the tool slices into the wood.

While Stocksdale works within a limited range of basic shapes, his repertoire of woods is far broader. Stocksdale estimates he has turned over 200 varieties; his wood scouts, scattered throughout the country, constantly watch for interesting material at lumberyards and other sources. One scout, an orthodontist and amateur turner, cut down 30 wild persimmon trees in Texas—a weed

BOB STOCKSDALE

Although basically the same shape, the grain of these four bowls creates a different character. From left to right: putumuju from Brazil (c. 1978), muninga from Africa (1982), ebony from Ceylon (1979) and mesquite burl from Texas (1980).

One of Stocksdale's wood scouts found a specimen of persimmon with interesting graining for this bowl (1978).

BOB STOCKSDALE

tree plaguing farmers in the Southwest—in search of the dramatic black-and-white pattern known to exist in the wood on rare occasions. His efforts yielded just enough material to make four bowls. A few tropical woods, such as teak from Thailand, come directly from a supplier in the country of origin, and exotic species are generally shipped to him in log form by a London wood merchant. It is the rare wood he admires most: snakewood from Surinam in South America; pink ivory wood from Zululand in southern Africa; thuya burl, which ranks among his favorite woods, from the Atlas

Mountains of Morocco. (Stocksdale acquired a piece of this scarce wood shortly before the last commercial lot was destroyed by fire.) Another of Stocksdale's prized logs came from one of the few mature specimens of Hawaiian kou to survive a devastating moth attack years ago. The tree had been surreptitiously chopped down by an amateur turner while the owner was away, but the owner was lucky enough to retrieve the wood. He shipped it to Sam Maloof who, in turn, sent a portion to his friend Stocksdale. In exchange for the small log, Stocksdale traded the owner a large bowl from its highly

figured wood.

Through his many years of experience, Stocksdale has become an expert in wood identification and understanding how different varieties respond to cutting tools and finishes. Fifteen to 20 tons of wood fill his garage, basement and backyard. Although the techniques of turning have become fairly routine, the species, grain pattern and shape of the stock vary constantly and keep the work challenging. "I never work the same wood twice in a row," he says. "It takes longer that way but makes it more interesting."

Even Stocksdale's early bowls, like
this one of English yew that he
turned for his mother in the early
1950s, have unusual grace.

BOB STOCKSDALE

Stocksdale's philosophy toward woodworking seems anachronistic by contemporary standards. Instead of proclaiming his bowls to be works of art, as many young craftsmen do, or his craft to be a way of life, as Nakashima does, Stocksdale views turning strictly as a profession. "I always started at 8 o'clock in the morning and I quit *precisely* at 5 o'clock—right in the middle of a bowl!" he declares. This, he believes, accounts for his longevity in the field. "Most young craftsmen knock themselves out working far too hard. If you maintain regular hours, you're always eager to get back and finish what you've started and begin something new."

Like all turners, Stocksdale is staunchly independent and always works alone. "I have a one-man shop and expect to keep it that way. My efficiency drops about 50 percent when someone else is in the shop." He differs, however, from many fellow craftsmen by his rate of production. While a turner like David Ellsworth may labor weeks over a single piece, Stocksdale can generate several pieces a day. "I average about one to two bowls a day, fewer with difficult woods, and up to 12 small plates—it depends on the wood." His prices reflect his speed: His work frequently lists below that of lesser-known craftsmen. "I want my friends to be able to afford my work," he often says. Bowls turned from rare woods, however, are priced accordingly and may run as much as 10 times those of more common species.

Stocksdale's midwestern background accounts for his pragmatic approach to his craft and his uncompromising work ethic. He was born in Warren, Indiana, in 1913 and raised on his family's 185-acre farm, where he became well acquainted with machinery and tools. As a teenager, he repaired furniture and built antique reproductions. His first lathe was powered by a gasoline motor transplanted from a washing machine— the farm had no electricity. He used it solely to turn parts for his furniture, and was entirely self-taught. Bowls would come later.

With the outbreak of World War II, Stocksdale joined the 18,000 Americans who declared themselves conscientious

Three bowls of various shapes: from left to right, para kingwood from Brazil (1981) with a ring of light sapwood along the rim; rosewood from Madagascar (1979); and pau ferro from Brazil (1982).

A bowl of tulipwood (1985).

BOB STOCKSDALE

This spherical bowl (1982) is made
of African blackwood.

BOB STOCKSDALE

These bowls are turned from Madagascar ebony from East India (1982).

objectors. As a result, he spent nearly four years in three civilian public-service camps. "It didn't seem particularly unusual at the time. My objections were political and not religious like most of the others'. I simply believed wars never solved anything, and that the best way to stop war was not to participate."

Like Nakashima, internment could not repress Stocksdale's desire to learn and create. With the permission of camp officials, he arranged for his

lathe and tools to be shipped to him. Turning, he reasoned, would be an appropriate activity for the idle periods between assignments fighting forest fires and repairing machinery, and it also generated a modest income. Other internees became interested in his lathe and many learned the craft from him. During his last months of incarceration—spent in a camp in northern California's Feather River Canyon—he was granted furloughs to restore furniture for an antique dealer

in Berkeley, California.

Upon his release in 1946, Stocksdale purchased more tools and a better lathe, and set up shop in the basement of a house in Berkeley, which he bought with two friends from the camp. From that point on, he made his living with the lathe, creating wooden bowls, platters and other functional and decorative turned objects. Soon exclusive department stores, notably Gump's in San Francisco and Fraser's in Berkeley, carried his work and he was selling all he could produce. Stocksdale lives in that same house with his wife, Kay Sekimachi, a noted fiber artist whom he married in 1972. He has two children, Jay and Kim, by a previous marriage.

Participants in the many seminars and demonstrations Stocksdale conducts around the country are often shocked to learn he creates his magnificent bowls and platters with just a few basic machines and only one or two hand tools—no complicated technology or secret methods. Over 90 percent of his turning is performed on a lathe with a 12-inch swing (the diameter of the work the lathe will take), which he adapted to handle stock up to 18 inches in diameter. In addition to another lathe having a 12-inch swing, Stocksdale's basement workshop contains a third lathe made for him by a friend from I-beam girders. This lathe allows him to turn platters and salad bowls up to 31 inches in diameter. Stocksdale produces every shape in his repertoire using ordinary ½-inch-wide and 1-inch-wide

With its relatively plain profile, this bowl of she oak (c. 1960) represents an early stage in the evolution of Stocksdale's work.

BOB STOCKSDALE

A bowl of blackwood acacia from California (c. 1979).

gouges instead of standard long-and-strong turning tools. "I started turning on my own, which is why I've always used the gouge," he says. "It wasn't until later that I learned the books say you should use scraping tools."

Because wet wood is easier and faster to turn, Stocksdale prefers to work with green wood. "The wetter the better," is his rule. To retard the drying process, Stocksdale paints the ends of the wood to seal in the moisture—logs and slabs, sections of stump and branches, all serve as his raw materials. After sawing the wood to manageable size, Stocksdale scribes one or more circles on it with dividers to plan its use, then bandsaws out the rough blank.

A hole bored in the center of the blank with a standard drill press allows it to be screwed to the faceplate of the lathe for rough-turning of the outside of the bowl. To rough-turn the inside, he mounts the blank with a three-jaw chuck, or, for larger work (like salad bowls), screws the blank to the faceplate. Later, any holes he has made are drilled out and filled with ¼-inch plugs of the same wood.

When rough-shaping is finished (and the walls are approximately $^3/_8$ inch thick), Stocksdale dries the bowls over the kitchen range for a month before final-shaping the walls. These may be turned to a thickness of $^1/_8$ inch or less. Any cracks are filled with a mixture of epoxy and sawdust. Finally, with the lathe rotating at its slowest speed, he sands the piece, first with an electric disk sander, and then by hand using four to five successively finer grits of garnet paper.

How Stocksdale finishes the work depends upon its function and the wood. Most decorative bowls are lacquered, although a few are oiled. In the case of objects made from dense, nonporous woods such as lignum vitae, African blackwood and boxwood, Stocksdale leaves the wood unfinished to retain its natural luster.

As the interest in wood turning grew during the 1970s, Stocksdale was invited by crafts schools and woodworking organizations to lead seminars and demonstrations throughout the country and England. His work, which sells at fine craft galleries, has been included in many group exhibitions such as "Good Design" at New York's Museum of Modern Art; "Triennale" in Milan, Italy; the 1958 Brussels World's Fair; "California Design"—shows 8 through 11 at the Pasadena Art Museum; and "The Art of Wood Turning" at the American Craft Museum II. The Oakland Museum, Museum of Contemporary Crafts and several private galleries have sponsored one-man exhibitions of his work, and his bowls are in the permanent collections of the Boston Museum of Fine Arts, the Royal Scottish Museum, the Philadelphia Museum of Art, the American Craft Museum, the Long Beach Museum of Art and others. In 1978, the American Craft Council elected Stocksdale a fellow.

Bob Stocksdale provides a rare example of how a craftsman dedicated to a single objective can achieve the culmination of his craft. Without relying on gimmicks or novelties, he rescued the craft of turning from near extinction and gave it an eloquence and poetry previously unimagined.

This wildly figured bowl (1981) was turned from a rare specimen of thuya burl from Morocco—one of Stocksdale's favorite woods. The bowl's sloping form, often called a bird's mouth, results from the natural curvature of the log.

TAGE FRID

Frid designed this corner chair to fit a round table.

Tage Frid stands for the principle that furniture design must be based on a firm understanding of established techniques. "Design around construction" is the underlying theme of his lectures, articles and books. It was a lonely position to defend years ago when the craft scene was dominated by artistic concerns, but Frid persevered. Today the craftsmen he has taught, many of whom have become teachers themselves, number in the hundreds. As Jere Osgood, a former student and associate professor at the Swain School of Design's Program in Artisanry, puts it: "It's reached a point where most of the people teaching in wood programs on the East Coast are either students of Frid or students of his students. We in turn pass on his ideas, consciously and unconsciously, to our own students." Without a doubt, Frid represents the single greatest influence on American woodworking education today.

The American Craft Council recruited Frid from his native Denmark in 1948 to help establish the wood department of its School for American Craftsmen at Alfred University in New York—the country's first college-level program for training designers and craftsmen in wood. At that time, student craftsmen were still copying stock designs from popular do-it-yourself magazines and using the most rudimentary techniques. The early 1960s would see experimentation with form as art students began to find their way into the woodshop, but there was precious little technical expertise to back them up. The arrival of Tage Frid in America signaled not only a move to develop educational programs in wood-working, but it coincided with a growing sophistication of the craft as a whole.

The product of a typical Danish craft education, which included a traditional apprenticeship and modern university education, Frid brought to America his wealth of experience, an infectious enthusiasm for his craft and a subtle approach to design. For Frid believes that wood's innate beauty requires sensitivity and restraint, distinguishing it from other media, which force the craftsman to impose himself on the material. "It is not like metal," he has said, "a piece of metal by itself is very cold and has to be hammered, shaped and polished before people will even look at it. A piece of clay, which is really dirt, must be shaped, fired and glazed. But take a piece of wood—plane, sand and oil it, and you will find it is a beautiful thing. The more you do to it from then on, the more chance that you will make it worse. Therefore, working with a material of such natural beauty, I feel that we have to design very quietly and use simple forms."

Implementing his philosophy, Frid's own furniture is masterfully constructed and of plain, straightforward design. For instance, the liquor cabinet he made

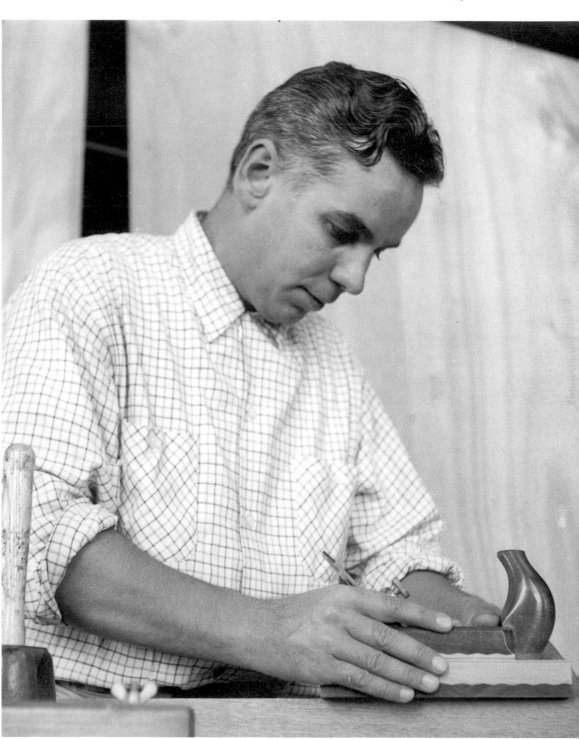

Frid in 1952—four years after he was recruited from his native Denmark to teach at the School for American Craftsmen.

in 1953 for "Designer Craftsmen: USA"—one of the first contemporary craft exhibitions in America—is a technical *tour de force*. The veneer is meticulously matched so the distinctive rosewood grain flows down the doors, over the carcase frame and drawer fronts, and down the legs. The inside is veneered with the same care as the outside. Few cabinetmakers in the country could handle such a difficult piece, yet the form could hardly be simpler.

A bit of design advice Frid gives his students is to draw ideas from various sources. "If you're going to steal, steal something good," he says. For example, he is fond of incorporating raised panels, a detail he borrowed from old doors, into the side and front panels of his credenzas and pedestal sideboards.

He believes the independent craftsman enjoys several advantages over mass production, including an ability to customize his work. Woodworkers, he stresses, should design with a sensitivity toward their clients' needs. "Furniture should be in proportion to the size of the buyer and reflect his or her personality. I don't think that anything can make a small person look more ridiculous—and perhaps make him feel smaller—than disappearing into an oversized upholstered chair." Small shops, he points out, also carry a lower over-

Regarding this liquor cabinet (1953), Frid said, "when you book-match veneer you only have one try." The rich rosewood grain continues up the legs and drawers, along the lip of the base and along the doors. The veneer on the inside of the doors matches the pattern on the outside.

head than industry because they bypass middlemen who mark up the price of the work.

Like his cabinets, Frid's chairs are unpretentious and superbly crafted. Here, his standards are particularly high. "There are certain requirements a chair must have when you design it. It should be designed so that it looks inviting to sit on, and when you do sit on it, the chair legs should not spread. You should be able to sit in it in various positions, and it should be able to take the weight of a person under stress circumstances."

Over the years, Frid refined a series of dining chairs he calls Chair #1, #2 and #3. Chair #1 was developed in 1951 for the conference room of the American Craft Council. He redesigned it in 1979 for a commission from the Museum of Fine Arts in Boston to create seating for their Greek wing and designated it Chair #2. Inspired by the furniture he saw on ancient Greek vases, he eliminated the stretcher of Chair #1 and increased the sweep of the back legs using lamination. Despite the origin of the design, Chair #2 has a decidedly contemporary feel. Recently, Frid changed the chair again by tapering the seat to make it lighter and simpler, and called it Chair #3.

His nylon-cord rocker (see p. 61) contains no wooden joinery except for a few shallow tenons for alignment.

Frid created this cherry chair, called Chair #2, in 1979—the second in an evolution of side chairs. It was commissioned for the Greek wing of the Boston Museum of Fine Arts. The design of the chair is simpler and more graceful than its predecessor, Chair #1 (1951), at right.

Frid's famous three-legged stools
are the epitomy of economy. Their
careful design makes them remarkably sturdy and comfortable. The
joints are round tenons. This group
is walnut.

TAGE FRID

The sides are pieces of ¼-inch-thick aluminum sandwiched between pieces of walnut and locked together with machine bolts. The nylon parachute cord strung across the back and seat conforms to the body in any position. The design of the frame, with the back and seat support emanating from the same point on the rocker, was inspired by the petals of a tulip.

Frid had always disliked three-legged chairs and stools where the odd leg is in the back, because they tip over when you lean to the side. But an idea for a three-legged stool of exactly that configuration came to him while seated on a fence watching his horses. "After sitting there for three or four hours, I suddenly realized it didn't hurt. Then I did some calculations and studied fannies for a while, and I found that a curved piece just 16 inches long and six inches wide was all the support you needed." The result is the quintessence of economy. Each stool consists of a short seat and narrow upright just high enough to support the lower back. No matter how you lean, you cannot tip it over.

Frid solved the problem of arranging chairs around a circular dining table by designing a corner chair (shown on p. 48) that fits together with its mates on an angle "like slices of a pie." Except for the way the sitter straddles the front corner, the design is reminiscent of the furniture of his native Denmark, particularly the work of the renowned architect-designer Hans J. Wegner, for whom Fried once worked.

In his nearly 60 years in the craft, Frid has continuously strived to broaden and hone his skills by ac-

cepting commissions for commercial, religious and domestic interiors. He finds satisfaction in the challenge afforded by each new situation. Unlike some craftsmen who see kitchen cabinets as drudgery, Frid believes they offer woodworkers some of the most stimulating and demanding assignments. "The kitchen is where you do most of your living, why should it be any less beautiful than the living room? I take kitchens seriously, and try to create a natural flow so the components are easy and inviting to use."

For the kitchen of Charles and Louise Hallick, Frid built deep drawers that open without the use of hardware. The layout is designed so as not to

Frid holds kitchens in high esteem, calling them the center of every home and a challenge for the woodworker. Each feature of this kitchen, constructed for Charles and Louise Hallick of Rochester, New York, was designed to be functional. The drawers, for example, slide on ball bearings, permitting easy access to the back. Instead of handles, Frid used handholds.

In a kitchen for the home of Janice and Paul Roman, publishers of *Fine Woodworking* magazine, Frid built cherry cabinets with tamboured doors (1984). The cabinets taper inward at the bottom so you don't hit your head when you lean over.

interfere with what Frid calls the "invisible walks." More recently, he designed and constructed the kitchen of Paul and Janice Roman, publishers of *Fine Woodworking* magazine. All the drawers slide on ball bearings so they can be completely extended for access to items stored in the back. The upper cabinets have tamboured doors and taper at the bottom so you won't bump your head when you lean over.

Born in Copenhagen in 1915, the son of a silversmith, it was young Frid's job to help polish the silver. By age 13, he knew he wanted nothing to do with silversmithing. Because he was a poor student, he says his parents suggested he apprentice to a cabinetmaker named Gronlund Jensen. Not knowing what else to do, he agreed.

The apprentice's lot has always been a hard one. The hours and work were grueling, even by adult standards, and corporeal punishment was practically mandatory. "The first couple of years I hated every minute and was planning to run away, which, thank heavens, never happened," says Frid. After working 10-hour days, six days a week in the shop, he took classes at night in mechanical drawing at a technical school. Yet the arrangement had many advantages over the American educational system. Master and apprentice signed a five-year contract. The appren-

Frid has made three versions of this sideboard, which is secured to the wall and supported by an angular, tapering pedestal. He says he borrowed the raised pattern on the drawer fronts from old doors. This cabinet in walnut (1974) is the first of the series.

TAGE FRID

With its clean curved lines, Frid's walnut and redwood grandmother clock (1984) betrays his Danish background. Frid created the swollen front of the case with a hand router. He selected redwood burl for the face because "I didn't want a white face staring back at me all day."

The fiddleback Honduras mahogany top of this table (1982) rotates 90° and flips open to twice its size.

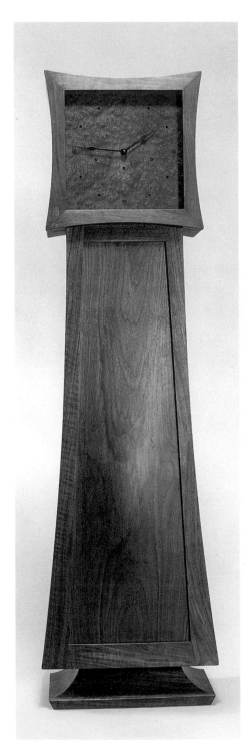

tice received a nominal wage and the master paid for his tuition at night school. If after five years the apprentice could not demonstrate the necessary skills of his trade and the master was found to be at fault, the master was legally bound to pay the apprentice a journeyman's salary while he continued his training in another shop.

A panel of experts evaluated the apprentice's proficiency. First, the apprentice made scale drawings of a major piece, which, if deemed acceptable, he would then build. At any time, the judges could enter the shop without warning to monitor the work's progress. When it was finished, the piece was displayed at the town hall, where officials scrutinized it with mirrors on wheels to examine the underside. They would turn the drawers upside-down and switch them around to make certain they fitted exactly. "I didn't learn much about designing," says Frid, "but when I finished I had an excellent understanding of wood as a material; its strengths and limits, and how to put furniture together."

After completing his apprenticeship, Frid continued his education by working in a variety of capacities including constructing the interiors of luxury liners. "It was the hardest job I ever had," he says, "because there wasn't a straight line anywhere. One window took almost an entire day." Eventually he decided to go to college and studied at the Vedins School and School of Interior Design in Copenhagen, from which he graduated in 1944.

After World War II, Frid moved to Reykjavik, Iceland, where he designed

interiors and married Emma Jacobsen in 1946. "I got married in Iceland," he jokes, "because I thought the wedding night would be six months long." They returned to Denmark in 1947 for the birth of their son.

The American Craft Council learned of Frid through a Danish arts-and-crafts publication and invited him to teach in the wood program of the School for American Craftsmen, then located at Alfred University in New York. His reception, he recalls, was less than cordial. The students were primarily interested in self-expression and ignored Frid's message concerning the importance of technique. Frid recalls, "when I arrived at the school, the students and some of the teachers kept talking about the 'freedom of the material.' This sounded interesting to me and I could hardly wait to find out what it was all about. It didn't take me very long to find out when I started teaching. They did not have control of the material, so many of the things made were actually accidents . . . I couldn't believe how far behind they were in construction and design. I kept telling them what they were doing wrong, and no one would talk to me."

Frustrated by his inability to reach his students and painfully aware of their cool response, Frid thought he would be fired. "So I figured I'd earn some money before they sent me back to Denmark. I took a mahogany plank from the school's woodpile and began making a coffee table." No longer concerned with explaining his ideas, he worked at his natural pace. Soon the students noticed what he was doing

and gathered around his workbench, amazed by his speed and precision. Seeing the master craftsman at work, they learned more than he was able to communicate over the previous six months. His affiliation with the school lasted for 14 years, including the move to Rochester, New York. In 1962, Frid was recruited to head the department of woodworking and furniture design at the Rhode Island School of Design in Providence, where he taught until his retirement in 1985.

After the School for American Craftsmen moved to Rochester, Frid joined with other faculty members in 1951 to establish Shop One, one of the first fine craft galleries owned and operated by craftsmen. He also served as designer and consultant to several businesses that produced furniture and accessories. His own company, Donovan and Frid, created interiors and custom furniture.

In 1975, Paul Roman persuaded Frid to join the editorial staff of *Fine Woodworking* magazine. Since then, Frid has written dozens of technical articles, which later became the basis for three books, *Tage Frid Teaches Woodworking—Joinery: Tools and Techniques* (1979), *Tage Frid Teaches Woodworking—Shaping, Veneering, Finishing* (1981), and a third volume on design and construction (published in 1985). His work appears in the collections of the Boston Museum of Fine Arts and the Rhode Island School of Design's Museum of Art. In 1985, Frid was elected a Fellow by the American Craft Council.

Frid continues to be a favorite speaker on the lecture circuit and fre-

This turned bowl and lid of cherry (c. 1966) has a stone handle that was cut by Frid's wife, Emma, an amateur lapidary.

Notable for its absence of joinery,
Frid used six bolts and bell nuts to
fasten the walnut shell of this
rocker (1984) to an aluminum core.
The seat is strung with nylon parachute cord.

TAGE FRID

quently gives demonstrations and seminars at craft schools and tool outlets, engaging his audiences with his expertise and charming them with his warm sense of humor. He has also exhibited at and judged many exhibitions. The Frids live in Foster, Rhode Island, in a 200-year-old farmhouse. His spacious workshop is located in a barn out back. They have two children, Peter and Ann, and a grandson, also named Tage.

Many pursue teaching as an alterna-

tive to the commercial world. Frid has proven he can succeed in both. To him, teaching represents another opportunity to create, only instead of wood, the creative talents of his students serve as his raw materials. "I love to teach because I learn so much," he says. "Students are always asking me 'why can't I do it this way?' Occasionally, it's something I never thought of and, as long as it doesn't break the rules of the material, I answer 'why not?'"

This cherry bench with leather seat is over 4 ft. long and nearly 2 ft. in depth.

A detail of a serving cart (1984)
showing the wooden wheel formed
by a technique called bricklaying.
The cart has a carving board built
into the top.

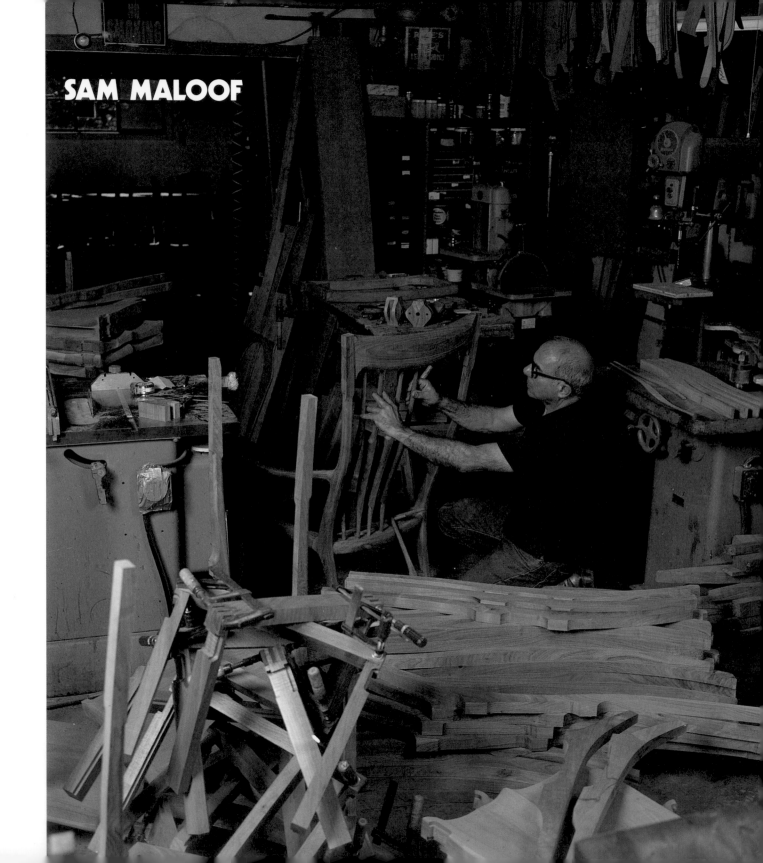

SAM MALOOF

As one of American woodworking's eminent practitioners, Sam Maloof has proven that a furnituremaker can succeed simply by producing fine work without resorting to novelty or mass production. Instead of feeling compelled to introduce new forms, Maloof concentrates on refining and improving his existing designs. And there is no mistaking the Maloof style—simple, rounded parts that flow together, reflecting his concern and involvement with every detail of his work—it pervades everything he makes. "People ask me why I do not go off on a tangent and work in different directions. My answer is that I have not really perfected what I am doing now. I do not think I ever will. Every commission I receive remains a challenge . . . I do not think you have to change just for the sake of change. If the piece is good, it's good. Ten different chairs can evolve from one design." This philosophy, also practiced by his friend Bob Stocksdale, represents the traditional approach toward the craft, which emphasizes making as much as designing.

To many people, the name Maloof is synonymous with chairs—considered by many craftsmen to be the most difficult piece of furniture to make. While Maloof makes a wide variety of furniture, including desks, tables and chests, he prefers the challenge of designing a comfortable and durable chair that is attractive as well. "I want my chairs to invite you to sit on them, to embrace you. If it doesn't sit well, it's a bad chair." Maloof uses his own body as a standard—when cutting slats for the chair backs, he saws them by eye and tests them against his back, adjusting the curve until it feels right.

Designing, he insists, is an instinctive process that cannot be taught. "My approach to solving many structural and design problems is to rely on my common sense and experience. I simply make decisions by eye. I use my forefinger and thumb for calipers and let my eye and intuition and years of experience do the rest."

Maloof's ability to vary a theme is an exciting facet of his work. To him, it is a more mature and less self-indulgent skill than developing new and exotic designs. His rocking chairs illustrate this talent: By altering the shape of the arms or the curve of the spindles, he creates several different chairs from one basic shape. Rockers are by far his most popular designs. (See pp. 68 and 69.) The Museum of Fine Arts in Boston, which boasts one of the country's finest collections of antique and comtemporary American furniture, selected a walnut rocker for their prestigious publication *Masterpieces from the Boston Museum.*

In Maloof's favorite design, his low-back dining chair (shown on facing page), the arms sweep so low—almost to the seat—that they serve as handles instead of armrests. The backrest provides good low-back support. The low-back reclining is one of his most sculptural works—the lines flow continuously throughout without ever becoming ill-defined or fuzzy. Maloof makes a graceful two-seat settee in the same style. Another chair, with similar

This spindly chair (1950) is an early version of the low-back chair first developed in 1948. Over the years Maloof softened the joints and shaped the parts until it has become one of his most sculptural pieces.

Joinery is an integral part of the design in Maloof's work, as in this chair where the leg flows into the seat and arm.

arms but a high spindle back, is known as the Texas chair because of its distinctive horned crest (see p. 81).

There are approximately 10 basic chairs in Maloof's repertoire, from which he produces about 50 designs. The backs may be high or low, with or without spindles, curved or straight. Within a particular design the variations become increasingly more subtle, depending on how Maloof sculpts the joints or varies the lines. Most chairs can be ordered with upholstery, which he subcontracts to a professional upholsterer—Maloof prefers plain fabrics in solid colors. Though the combinations are endless, Maloof is sensitive to suggestions that he constructs his chairs from stock parts. "I *do not* make up a bunch of parts and reach into bins to assemble a chair from pieces: I do not stockpile any of my pieces; each piece is made for the person who orders it."

Maloof's other pieces of furniture exhibit the same elegant proportions and attention to detail as his chairs. His print stand and music rack have clever devices for adjusting height and angle that are made entirely from wood. His well-known cradle hutch was created for a 1966 exhibition at the American Craft Museum called "The Bed." His daughter-in-law was pregnant at the time and he planned to make her a cradle, so he began, as most furnituremakers do, by considering the piece's function. "I thought, 'Well, most cradles are so low that a mother would break her back putting the baby in. Why not bring it up high?' And I thought then, 'Why not make drawers under it for the baby's clothes?' I started on that idea and then I thought, 'Well, you know, why not enclose it completely in a cabinet with storage space for blankets up above?' And this is the way the thing is conceived.'' The latest version includes a pull-out shelf for diapering. Regardless of Maloof's backlog, he postpones all other work to complete an order for a cradle hutch so it will be ready in time for the baby.

Maloof's desks and tables are simple and well-crafted. Most are supported by a pedestal or trestle arrangement (see p. 80). Says Maloof, ''A simple object is the most difficult to design and the most difficult to make. Attempts to be 'different' result in mere cleverness and have little value. To be original just to be original leads nowhere; the design fails to transcend the woodworker's ego.''

Some young woodworkers find Maloof's emphasis on perfecting the same designs to be stagnant, and a few dismiss his style as a rehash of Danish modern. He defends the similarity between his work and Scandinavian furniture as coincidental and points out that he established his style years before he saw the Danish imports. ''Today, many woodworkers have made a big splash by resurrecting art deco and calling it post-modern,'' he says. ''To me, it's just another trend.'' Maloof has seen many trends come and go during

In searching for a way to save new mothers from bending over to lift their babies out of the cradle, Maloof developed this cradle hutch. Cabinets and drawers provide storage and a pull-out shelf is a handy place to change diapers.

SAM MALOOF

Nursing mothers find this rocker, with curved spindles and scooped arms, especially comfortable. It is currently on loan to the Philadelphia Museum of Art.

his long career. He believes the more restrained and unaffected the design, the broader its appeal.

Maloof's customers gladly wait years for their orders. And he describes his clients—who now number in the thousands—as friends. Maloof enjoys this relationship and feels it distinguishes handmade objects from industrially produced ones. "There is a mutual understanding between the maker of an object and the user as there is a communication between the craftsman and the material he works with. . . . This becomes a triple communication when it embraces the person who has commissioned the piece." He prefers dealing directly with customers rather than through galleries or decorators. When-

ever possible, he gets to know clients personally and visits their homes.

Maloof typifies the first generation of postwar craftsmen who were self-taught, fiercely independent and forced to create their own market at a time when handmade furniture was nearly unknown. Born in the farming town of Chino, California, in 1916, Maloof was the seventh of nine children. His Lebanese parents taught him self-reliance, respect for other cultures and a deep love of family. He also inherited their vision that anything is possible in America for those who persevere. As a child, Maloof showed an interest in drawing and making things—he carved most of his own toys and when he was in high school, he constructed furniture

A rocker with straight spindles

SAM MALOOF

In assembling the spiral staircase, Maloof used nearly 100 clamps. The rail is laminated from six separate pieces and bent.

Maloof finished this spiral staircase for his wife's office in 1983. The wood came from discarded crates used to ship motorboats from the Far East.

for his family. Winning a high-school art contest led to his first job as a graphic artist.

His drafting skills helped him become the Army's youngest master sergeant in 1942. Following his discharge in 1945, he worked as a commercial artist for a decal company in Los Angeles, California, which still produces his designs. To furnish his first apartment, he enrolled in night school to use the machinery and built his own furniture. Like drawing, Maloof learned woodworking on his own. "I had no formal training of any kind . . . I just did things the way I thought they ought to be done." In 1947, he served as a studio assistant to artist and teacher Millard Sheets, who exposed him to a vari-

SAM MALOOF

ety of media including sculpture, painting, murals, mosaics and pottery.

While working for Sheets, Maloof met Alfreda Louise Ward, an art student and former arts-and-crafts director of the Indian school in Santa Fe, New Mexico. They married in 1948. Maloof describes their relationship as a partnership, and considers the support of one's spouse essential to a successful career in crafts. "Often young couples come to my shop to seek advice about getting started . . . I tell them that both partners must be equally committed because, very bluntly, the kind of money that a woodworker earns involves hardship and sacrifice. The rewards of woodworking are not monetary." Alfreda Maloof manages the books and handles the correspondence. They have two children, Slimen and Marilou, and four grandchildren. Slimen worked in his father's shop since the age of 10 and later opened his own studio on the family compound.

Shortly after he was married, Maloof quit his job with Sheets for a career making furniture while working freelance as a graphic artist. It was difficult

These print racks have wooden hinges and a wooden mechanism to adjust the height and angle of the leaves.

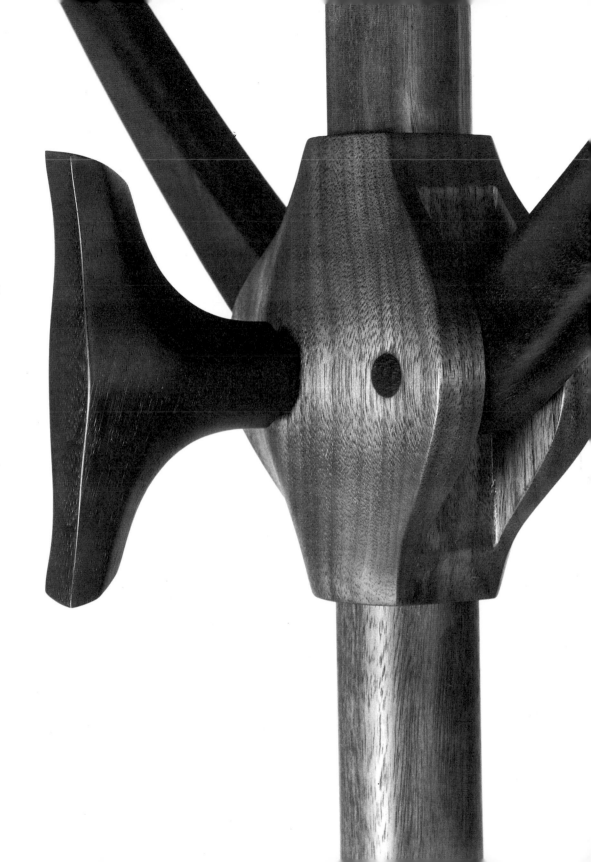

giving up a secure job at a time when furnituremakers who made their own designs were rare. "If it was material security that I sought," he says, "I would not have become a woodworker . . . .To me the spiritual satisfaction my work has brought me, coupled with the satisfaction it brings those for whom I have done the work outweighs any other benefit I might gain."

Although the basic forms of his furniture haven't changed since the 1950s, his early designs were more angular and less organic than his recent work. The prototype of the low-back chair, made in 1950, appears gangly and awkward compared to its present form. Over the years, Maloof refined the chair by softening the joints, curving the back leg and arms and sculpting the back and seat (see pp. 64 and 65).

When Maloof first married, he couldn't afford to keep any of the furniture he made. One day, a friend gave him some encrusted plywood that had been used for cement forms. After having it sandblasted, Maloof constructed tables and cabinets from it for the dining and living rooms (see p. 76). Through the contact of another friend, Maloof's home was featured in *Better Homes and Gardens* in 1951 as an example of a tract-house interior. Upon seeing his furniture, the magazine asked him to make scale drawings to market to amateur woodworkers. Based on that

A detail of the print rack showing the wooden device for adjusting the leaves. The pins are brass, plugged with ebony. The knob was roughsawn on the bandsaw, then shaped and finished by hand.

publicity and encouragement from his friends and clients, Maloof stopped the freelance work for a full-time career in woodworking. He was 32.

His first commission was a financial disaster. The order was for a dining table, 10 chairs and a buffet, and he wanted everything to be perfect. To test the joinery, he made a sample chair and dropped it from the roof of his house. The wood snapped but the joints held firmly, so he confidently completed the order. But on the instruction of the customer's interior decorator, and against his own better judgment, he stained the wood a murky gray. After installing the furniture in the client's living room, she told him she loved the designs but hated the color. Naive and inexperienced, he carted all the furniture back home and refinished it at his own expense.

Despite its inauspicious beginning, Maloof persisted at his new career. Gradually, his reputation grew to where he could attract sufficient business without relying on decorators or galleries. In 1953, he and his wife purchased three acres of land in a lemon grove in Alta Loma, California, near the San Gabriel Mountains, about 50 miles from Los Angeles. They lived in a flimsy shack and Maloof set up his shop in

Maloof's workshop connects to the house, allowing him to work late into the night.

a chicken coop with a dirt floor. Fulfilling many a woodworker's dream, he began to build his own home and studio.

Construction was slow and thoughtful. When he finished one room, he built another, more complex one, next to it. The redwood house progresses like blocks around a central courtyard filled with potted plants. Like the homes built by his contemporaries Esherick and Carpenter, every detail is crafted to reflect Maloof's sense of design and passion for wood. Maloof carved a different latch for each door and shaped the frame of every window. A spiral staircase (see pp. 70 and 71) is the latest major addition and, typical of his style, the joints are sculpted so the steps taper from the center post like petals on a flower. The laminated handrail is supported by uprights that bow outward like the spindles of his chairs. The railing ends in a delicately carved knob indicative of the subtle flourishes he enjoys incorporating in larger pieces. The Maloofs purchased three additional acres in 1955, which they operate as a working lemon grove. The fragrance of lemon blossoms permeates the house and workshop.

Their extensive collection of contemporary and pre-Columbian pottery, sculpture, Indian art, paintings and textiles lines the shelves and beams

Before he could afford to keep any of the furniture he made, Maloof furnished his home in 1948 with plywood furniture of his own design. The furniture was featured in *Better Homes and Gardens* magazine in 1951.

throughout the house. They also collect wood turnings by Bob Stocksdale, Ed Moulthrop, David Ellsworth and others. Samples of Maloof's furniture fill every room, and potential clients stroll through the house to view his designs. His wife accepts the traffic as part of her husband's business: "Even if we had a showroom," she says, "Sam would still invite them into the house."

Because there were few role models available to Maloof when he began his career, he took it upon himself to become a leader of craft causes. His initial contact with other designer-woodworkers occurred in 1957 at the first American Craft Conference held at Asilomar, California, where he met Wharton Esherick, Walker Weed, Bob Stocksdale, Arthur Espenet Carpenter, John Kapell and Tage Frid. Recalls Maloof, "What we all had in common was that we were doing what we wanted to do. None of us was a conformist. None of us wanted to be tied up or bound. I believe all of us were seeking spiritual well-being in what we were doing. We were not using our work as a means of avoiding responsibility in a material sense."

His interest in craft organizations led him to help establish the Southern California Designer-Craftsmen in the mid-1950s, and he served as a craftsman-trustee of the American Craft Council from 1973 to 1976, and as a trustee since 1979. He was named a fellow of the American Craft Council in 1975. In 1959, he traveled to Iran and Lebanon as a consultant to a State Department project to encourage craft industries in developing nations. This

The interior of the Maloof home.

The handles of this chest of drawers are shaped from the drawer front itself. The excess material is used for the sides and backs of the drawers. Nothing is wasted. The case is secured with screws and the holes plugged with ebony.

gave him the opportunity to visit his parents' home town. The experience left a deep impression on him. Some of his Lebanese relatives, however, felt sorry for Maloof—they thought their American cousin couldn't possibly be successful if he worked with his hands. He participated in a similar project in 1963 in El Salvador. Maloof frequently lectures on his work and often serves on the juries of craft shows. He recently completed a book on his career titled *Sam Maloof: Woodworker,* which is lavishly illustrated with drawings and photographs of his work, and he is writing another. He has been the subject of two documentary films, and his furniture has won numerous awards. It is included in many prestigious collections, and in fact Maloof was the first contemporary furnituremaker to be included in the permanent collection of the White House. In 1985, he received a $650,000 grant from the prestigious McArthur Foundation.

Throughout most of his career, Maloof has employed one assistant and one apprentice or trainee. Their role in his shop is generally limited to sanding and finishing. Once he has shaped a piece's basic form, he will allow an assistant who has acquired a genuine feel for the work to complete the finish-shaping. But Maloof steadfastly refuses to relinquish any of the creative or technical processes—it is his way of maintaining quality and a personal

Similar versions of this low-back settee appear in the permanent collections of the Boston Museum of Fine Arts and the St. Louis Art Museum.

involvement with every piece. "Unfortunately for the apprentice, in my shop I do all the designing, all the layout, cutting, joinery and rough-shaping, as well as much of the finish-shaping and even some sanding. I can truthfully say that I have made everything that goes out of my shop. It may seem selfish, but if I let other people put my things together I couldn't call them my pieces. In some ways, this defeats the purpose of having an apprentice. But whenever I do something they haven't seen, I call them over and explain what I'm doing in hopes that when they leave they'll be able to do it on their own."

Maloof works at a feverish pace—60 to 70 hours a week—as only a man who loves his work can, and produces approximately 60 pieces a year. His

speed comes from years of experience and the efficient use of power tools. He is especially adept with the bandsaw and router, although he estimates 80 percent of his work is done by hand. He performs most machine processes—cuting out forms, boring holes and rough-shaping—freehand, and claims he can cut two matching parts to within $1/_{32}$ inch by eye.

SAM MALOOF

In contrast to woodworkers such as George Nakashima and James Krenov, who may hold a slab of wood aside for years while considering its perfect use, Maloof takes a more pragmatic approach. "Some woodworkers talk about the necessity of contemplating a piece of wood and letting it tell them how it wants to be used. This is fine, but time is precious. Personally, there are so many pieces of furniture for which I have mental drawings and there are so many more pieces of wood in my future that I have no time for leisurely conversations with a single piece. My communications with wood, therefore, are very efficiently condensed. . . I relate intensely to wood. The pieces that will become furniture are chosen with a mixture of common sense and love, and there is no reason for this process to be long and arduous."

Many consider Maloof a master of joinery, especially because of the wooden hinges and pedestal joinery in his tables, and his chair joinery. Rather than concealing joinery, Maloof incorporates it into the design. "If it's a good joint, why hide it?" he asks. He relies heavily on the router and power saws to create the joints, and on chisels, rasps and Surforms to sculpt them. But whether he uses a complex wooden joint or a simple metal screw, which he calls a "metal dowel," depends on which he judges to be sturdiest under the circumstances. He harbors no loyalty to traditional methods. Although woodworking purists may scoff at his use of dowels and screws, he feels they can be stronger than conventional joints because less wood is removed. He also notes that his work has withstood years of use without problems.

Ninety-five percent of Maloof's work is made of black walnut—he likes the way it works and feels its rich color

Maloof makes a variety of trestle desks, including this one with a file drawer.

and grain complement his designs. "Some woodworkers like to work in a wide variety of woods, but my palette is quite small. Furthermore, I do not mix my woods and do not use metal hardware of any kind or other materials for embellishment." By selecting lumber with knotholes, he obtains wood with more interesting grain than the expensive grades. Whenever possible, he accentuates the contrast between the walnut's honey-colored sapwood and its dark heartwood. He finishes his furniture with a mixture of varnish, boiled linseed oil, raw tung oil and beeswax, although he adds thinner and varnish for tabletops to make them water resistant.

Over the years Maloof has rejected lucrative offers to mass-produce his designs because he feels furniture suffers when the creative process is separated from the technical one. "I believe the best results are achieved when the designer is also the craftsman. The two should not be separated, because they share equal responsibility. It is difficult for the designer to transmit his ideas to someone else. Thus in the mass method the machine craftsman must take part in the designing, and as a result the designer often wonders if the finished product is actually what he intended. The designer-craftsman, however, works out his own design. He can make changes as he works. If a piece does not look right, he corrects it. He isn't designing for a machine but for an individual who seeks the finest quality of workmanship."

Called the Texas chair because of its distinctive horned crest, the arms of this spindle-back chair are so low they obviate the need for stretchers below.

ARTHUR ESPENET CARPENTER

Californian woodworking is often considered separate and distinct from the craft as it's practiced elsewhere in the country. Many characteristics commonly ascribed to West Coast furnituremaking—emphasis on form over technique, experimentation with new materials and an organic style of flowing lines—stem from the work of Arthur Espenet Carpenter. As one of the first designer-craftsmen to emerge after World War II, Carpenter's designs and independent life-style have become a model for the succeeding generation of woodworkers, especially in the San Francisco Bay area. Part of his influence on young craftsmen derives from his willingness to share his experiences. "Of all my peers, including Nakashima and Esherick, none of them seemed interested in teaching," Carpenter says. "I, however, see no exclusivity in this racket. I'm actually aggressive about bringing people into the fold." Hundreds of budding woodworkers have visited his studio to elicit advice on issues ranging from technical problems to financial concerns. Over the years, he has trained dozens of apprentices and served as a leading spirit of the Baulines Craftsman's Guild, a local multimedia craft organization.

Carpenter's work can best be defined by the five criteria he considers when designing furniture: function, durability,

Carpenter used a bandsaw to create the pigeonhole arrangement of this walnut and mutenye roll-top desk (1970). The tamboured top consists of ¾-in. by ¾-in. strips glued to a canvas back.

simplicity, sensuality and practicality of construction. Function is the starting point for all his work. "When I sit down with my clipboard in my lap fiddling over a new design, I shut out all references to furniture I've seen and concentrate on the functional requirements of the piece. First I draw the points or lines or angles that satisfy these specifications. Then I attempt to arrange the form and joinery in an unclichéd and aesthetically pleasing manner." As a maker of utilitarian objects, Carpenter insists his furniture withstand everyday wear. If a piece

The front of this teak bandsaw box is sculpted into drawer pulls (1972). There's even a drawer within a drawer on the left.

Calling it his most successful design, Carpenter says the Wishbone chair comes the closest to satisfying his standards of function, durability, simplicity, sensuality and practicality of construction. The legs are laminated and bent around a dry form. A wedge at the bottom splays the foot. The back is shaped by hand. The frame of the seat is lap-jointed and covered with saddle-leather webbing. Six ¼-in. bolts hold the prefinished parts together, which in this chair (1980) are cherry. The holes are countersunk and capped with walnut plugs and beeswax to permit access if adjustments or disassembly are necessary. The chair comes with optional laminated and steam-bent arms.

This holly bandsaw box (1970) is another of Carpenter's experiments. He bleached the holly, which is nearly white in its natural state, to extract every last bit of color. The knobs are ebony.

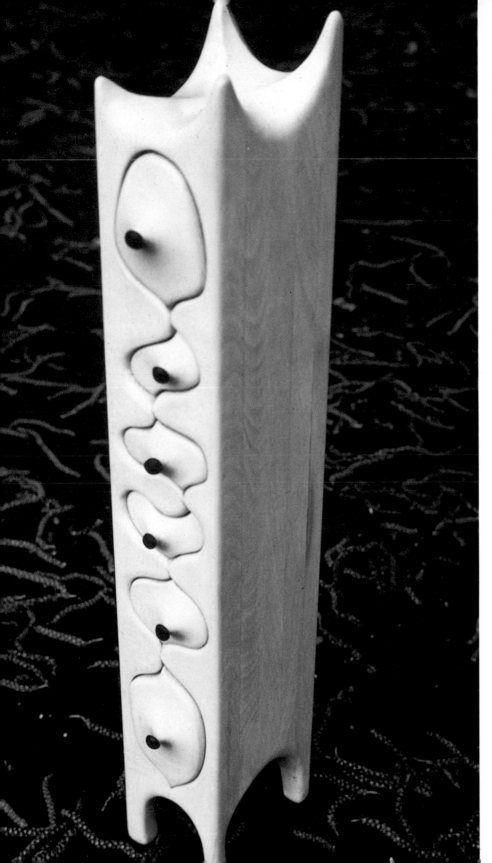

breaks under normal use, he repairs it without charge—even years later. As to simplicity and sensuality, Carpenter advises, ''Never use a compound curve when a simple sweep will do just as well.'' Carpenter pushed his philosophy of design to its limit with a line of furniture he built in 1966 for the Mill Valley Public Library. Besides having straightforward construction—he had to complete 100 pieces in 12 months with just one assistant—the work needed to be durable and economical. To fulfill these requirements, he developed a design he termed *Lu Pan,* after an ancient Chinese style, consisting of three flat surfaces joined at right angles to form and inverted ''U'' (p. 94).

Carpenter's Wishbone chair is the design he feels best satisfies his criteria of function, durability, style and ease of production. He generally makes about eight at a time. The legs are laminated and the back is cut out on a bandsaw and shaped with a router and pneumatic drum sanders. The parts are finished off by hand and then bolted together—the only wood joinery occurs in the seat. Carpenter takes particular pride in being able to produce each chair in just two-and-a-half days. As Carpenter says, ''Anyone can make a complicated chair. It's hard to create one that's simple in design and simple in manufacture.''

Carpenter's genius for technical improvisations is responsible for many of his most original designs. Although he claims his ideas spring from laziness, they actually represent an effort to balance artistic freedom with economic reality. For example, his bandsaw boxes

ARTHUR ESPENET CARPENTER

The interior of Carpenter's house is paneled in solid redwood. The ceiling beams and verticals are fir dyed black. Most of the furniture is walnut except for the built-in bookcases, which are redwood.

Carpenter spent seven years constructing this circular home for his family. It stands 40 ft. in diameter with 2,100 square feet of floor space. *Life* magazine featured the house in 1966 as a tribute to creative domestic architecture.

(now widely imitated by woodworkers across the country) resulted from Carpenter's frustration with the time it took to make a standard drawer with its four sides and bottom. So he devised a method to eliminate the joinery and give the drawer a freeform shape at the same time. He simply cuts out the rough drawer from a solid block of wood with a bandsaw, bandsaws out the inside of the drawer, then slides the drawer back into the block. These allow him to test new ideas on a small scale before undertaking larger work. "Bandsaw boxes are more disposable than full-size pieces," says Carpenter. "You can bleach it, dye it, paint it with pink polka dots. If it doesn't work, what have you lost? It gives you that freedom to experiment you don't have with a six-foot chest of drawers." In one experiment, he bleached the color from a piece of holly, leaving the wood bone white (see p. 85). He employs a similar technique of cutting freeform openings with a bandsaw to create the pigeonholes in his roll-top desks—each arrangement is unique (see p. 82).

ARTHUR ESPENET CARPENTER

Using an auto-body grinder to sculpt the steps, Carpenter constructed this 16-ft. spiral staircase in 1969 for the home of David Davies of Redbank, New Jersey. The steps are hyedua and the center post is oak.

Over the years Carpenter has developed a number of innovations to make the creative process more practical. These include a jig for cutting dovetails more accurately and quickly than by hand (varieties of which are now in common use among woodworkers), and a machine to extract the inside of a rough bowl in one piece, enabling him to make two or more bowls from the same block of wood. And to sculpt the steps of a 16-foot-tall spiral staircase Carpenter constructed in 1969 for the home of David L. Davies in Redbank, New Jersey, Carpenter used an auto-body grinder.

Carpenter freely borrows his design ideas from nature—a good example is his Scallopshell desk which was originally commissioned by Kay Sekimachi, Bob Stocksdale's wife. This desk also illustrates how an inventive technique can engender a new piece: The shell pattern on the sides sprang from Carpenter's experiments with laminating contrasting woods. To complete the effect, he added texture by sculpting each section into ridges.

Occasionally, Carpenter's use of natural forms is literal, as when he scatters mushroom-shaped handles of various sizes across the front of a cabinet. Usually, however, the images are more subtle. Edges, for instance, are generally rounded over—a style often termed the "California roundover" and frequently attributed to Carpenter and Maloof; the parts in Carpenter's furniture seem to grow together, as in the arms and legs of his captain's chair or in the twisting base of his arc-leg table, where the members intertwine like vines. "I find

Carpenter's furniture is strongly organic, as exemplified by this walnut arc-leg table (1968). The circular top is 50 in. in diameter.

ARTHUR ESPENET CARPENTER

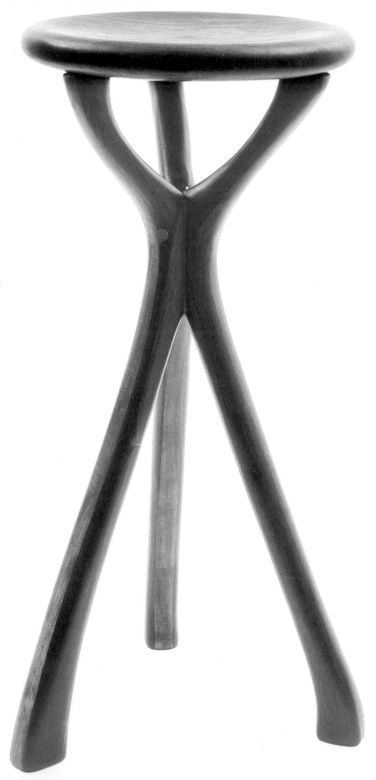

precision in surfaces to be deadening,"
he says, "like a note sounded without
harmonics or vibrato."

Unlike most woodworkers, Carpenter
uses scale models to develop new
designs. Their size (1:8) permits him to
mail the models to potential clients for
approval. Dozens of miniature cabinets,
chests, tables, chairs and desks line the
shelves of his showroom, testifying to
his versatility and tireless obsession
with trying new forms.

Rather than rely on former successes,
Carpenter's curiosity compels him to
experiment with new styles, techniques
and materials. "I think of myself as an
experimenter and originator. Maloof, for
instance, has a particular style and has
made it sing. . . . It's beautiful. But I
keep changing and revamping. I have
never made anything that I'm completely satisfied with."

Recently, his compulsion to inves-
tigate new areas has involved applying
color and materials such as plastic, steel
and enamel to his work. In his Mondrian chest, Carpenter faced the
drawers with brightly colored plastic
panels separated by black horizontal and
vertical lines, mimicking the style of
Dutch painter Piet Mondrian. Some
experiments seem playful, such as when
Carpenter paints a leg of his walnut
music stand kelly green, while others,
like the Mondrian chest, reveal a serious
interest in exploring new approaches to
color and form (see p. 97).

A rosewood stool (1967) that is
29 in. high.

Carpenter assembling a ten-sided
dome for a 1975 crafts fair.

ARTHUR ESPENET CARPENTER

Simplicity is the key to Carpenter's designs as in this hexagonal table/desk of American black walnut (1969). Like his *Lu Pan* designs, the sides are dovetailed to the top.

Although widely associated with California, Carpenter is actually a transplanted Easterner. Born in Brooklyn in 1920, his family later moved to Larchmont, New York, and Medford, Oregon, where he finished high school. Fulfilling his father's wishes, he studied economics at Dartmouth College, intending to become a corporate accountant. Following his graduation in 1942, he served four years in the South Pacific with naval intelligence and received four battle stars. "The four-year hiatus from economic demands and career obligations allowed me time to see that I didn't have to follow a given pattern and that perhaps I could go in my own direction, whatever that

would be," says Carpenter.

After the War, he worked as a stockboy and courier for his father's Oriental-art import business, where he gained an appreciation for finely made objects. "Oriental art taught me that craftsmanship was not a casual enterprise. . . . The Japanese have a saying to the effect that only the third generation in a family can attain true craftsmanship."

In 1947, Carpenter attended several design exhibitions at New York's Museum of Modern Art, which inspired him to alter his career plans. The exhibitions heralded the contribution of designers who brought contemporary styling to functional objects, and

This walnut and cherry double music stand (1968) appeared in "Woodenworks," the inaugural exhibition of the Smithsonian Institution's Renwick Gallery.

The pattern on the side of the Scallopshell desk originated from Carpenter's experiments with lamination. In this desk (1972), Carpenter inserted wedges of walnut between thin strips of lacewood to create the design and then sculpted each section to create the scallop.

included furniture by Marcel Bauer, Ludwig Mies van der Rohe, Charles Eames and Michael Thonet. "Those shows gave me the direction I was looking for. I was no artist—I was no craftsman, but even so I could make things that might be considered beautiful." So Carpenter quit his job and drove his 1938 Cadillac across the country to San Francisco because "it was as far from New York as I could get."

Using his modest savings and the $100 a month he received from the GI Bill to start a small business, Carpenter purchased a wood lathe and set up a shop in a condemned building on San Francisco's Mission Street. "I chose the lathe because it was the simplest tool

The pedestal dining table and captain's chair are two of Carpenter's most popular designs. This set, commissioned by a Sacramento television station in 1970, is walnut. The back of each chair is stack-laminated and the remainder is solid.

Carpenter used this simple design for a series of stools, benches, desks and tables he named *Lu Pan* after an ancient Chinese furniture style. Generally, each *Lu Pan* piece is composed of a single board with the sides dovetailed to the top so the grain continues across the top and down the sides.

for making the simplest object I could think of. In the forties there were no bowls from Denmark or Taiwan. Bob Stocksdale, so far as I knew, was the only bowlmaker who attempted more than the uncaring utility of Michigan kitchenware or the lacquered atrocities of redwood and bay sold to tourists on [California's] Highway 101. Later, I discovered the beautiful experimentations of James Prestini—many done prior to Stocksdale." It wasn't long before Carpenter realized he had underestimated the complexity of his new profession. Today, after nearly 40 years as a professional woodworker, Carpenter readily admits he's still learning the eccentricities of wood. "It's an

The Mayer sideboard was named after the buyer who commissioned the original prototype design—a common practice among craftsmen. This version in walnut was made in 1969.

ARTHUR ESPENET CARPENTER

obtrusive and obstinate material," he declares. "It wants to do its own thing. I try to select wood that is relatively plain—being more interested in line, form and color than its intrinsic beauty. Although, of course, you display pretty pieces in door panels or the tops of boxes."

His early experiences on the lathe provided him with a solid introduction to the properties of wood. "A lathe is a good place to start if you want to teach yourself how to handle materials of almost any consistency from balsa to brass. You control every process using just a few tools—it's a factory unto itself. . . . I gained a rapid education during those years on the lathe."

Within six months he sold his first bowl. After two years he was selling all he could produce, so he hired assistants to rough-out the bowls before he finished them. Edgar Kaufmann, Jr., former director of the department of industrial design at the Museum of Modern Art, saw Carpenter's bowls in a Chicago wholesale store and invited him to participate in the museum's "Good Design" shows from 1950 to 1954. "I had come full circle," recalls Carpenter, "back to the place and type of exhibit that gave me the impetus to make my move in the first place."

In search of greater artistic and financial flexibility, he expanded into furniture. "I started making coffee tables—simple things. I didn't know how to keep a board flat or what it was to dry out a piece of wood. At that time, I knew of no furniture designer-craftsmen in the Bay area. But there were plenty of old-time cabinetmakers

who knew how to treat wood and put things together, and they seemed delighted to have somebody they could teach. So I learned technique from them and did my own designing—never telling them how my experiments might subvert their techniques, which would have horrified them."

Carpenter's first furniture was primitive, what he called "slabs with turned sticks or wrought-iron bases. It was in the mid-fifties when I heard of and saw some of Esherick's work and began to see the possibilities of sculpting edges

and making things asymmetric." Besides Esherick, he identifies Sam Maloof, Bob Stocksdale and Danish furniture, which gained popularity in America in the 1950s, as early influences on his career.

Around 1954, a columnist for the *San Francisco Chronicle* dubbed him the "arty carpenter." The pun so appalled Carpenter that he adopted his middle name, Espenet, as his professional name.

When he began making furniture, he supplemented his income by entering the do-it-yourself market selling furni-

Carpenter began his woodworking career as a turner because he naively believed it was the easiest craft to master. These bowls of Honduras mahogany, Siamese teak and prima vera, 1953, represent examples of his work before he began creating furniture.

Carpenter created a series of chests inspired by the abstract paintings of Piet Mondrian. The case is walnut and the drawer fronts are madrone. The drawers are framed with strips of black plastic. An occasional drawer is faced with red or blue felt and plastic.

ARTHUR ESPENET CARPENTER

ture kits, wood and supplies. By 1958, the business was thriving and he had a staff of seven people. But his interest in "buying and selling palled as it had ten years before," says Carpenter, and he longed for a more creative challenge. "I found myself doing what I was trying to get away from when I started. So I took the opportunity of simultaneously moving out of San Francisco and dropping the do-it-yourself thing."

Using a small inheritance, he purchased a farm in 1957 in Bolinas, a secluded coastal town 20 miles north of San Francisco, and devoted himself to his furniture. He also set out to build a house. He wanted it to be practical, comfortable and attract publicity for his business. With plans by architects Robert Marquis and Claude Stoller, he constructed a magnificent circular home on a knoll overlooking the Bolinas Lagoon, doing all the work himself.

The layout of the house consists of an array of concentric and overlapping circles with the kitchen as the hub. One hundred and twenty sturdy fir timbers radiate like sunbeams from a ring of clerestory windows in the roof. A circular steel fireplace, designed and fabricated by Carpenter, stands within an arc-shaped, sunken area of the living room next to an indoor-outdoor pool once stocked with his son's collection of frogs and newts (see pp. 86 and 87).

The project took seven years to complete because he kept running out of money for materials. Much of the wood for the interior came from his workshop. The kitchen counter is a mosaic of walnut, oak, teak, cherry, maple and holly rippings. He originally intended to

construct the walls from plywood but just as the foundation was poured, a local lumbermill folded, enabling him to buy enough redwood boards at 5 cents a foot to panel the entire interior and exterior of the house.

Life magazine featured the house in 1966 as an example of creative domestic architecture. When the house was sold in a divorce settlement, Carpenter built a series of dome-shaped buildings near his workshop for new living quarters. Each room is a separate building and a different experiment in construction. The mild California

weather allows for a pleasant walk from the bedroom along a stone path to the kitchen.

Recognition of Carpenter's contribution to the craft came in 1972 when the Smithsonian Institution included Carpenter in the Renwick Gallery's inaugural exhibition along with Wharton Esherick, George Nakashima, Sam Maloof and Wendell Castle. Other museums that have shown his work include Boston's Institute of Contemporary Art, the Brooklyn Museum of Art, the Museum of Contemporary Crafts in New York, the Akron Art

Known as the Moon chair, this chair, constructed of stack laminated walnut, was commissioned in 1969 for a local chapel.

Carpenter faced the center drawer of this Peruvian walnut bandsaw box with a red enamel plate (1975). The drawers are lined with colored felt.

Carpenter experimented with mixing styles in this low stool by combining an organic top with angular asymmetric legs. The walnut top is left natural and the mahogany legs are dyed black to accentuate the contrast.

Institute, the Long Beach Museum of Art, Chicago's Museum of Science and Industry, the Oakland Museum and the Crocker Museum of Art in Sacramento. His work also appeared in "Objects: USA" and "California Design"—shows 8 through 11.

Carpenter has two children, both interested in crafts. His daughter, Victoria, is involved with ceramic sculpture and painting, and his son, Tripp, who learned to operate the lathe at the age of seven, assists his father in the shop.

Above all, Carpenter is a free spirit.

He was "doing his own thing" years before the cliché became popular. This independence affects his sense of design as well as his unconventional life-style. Unhampered by loyalty to a particular style or method, he constantly searches for new challenges and fresh approaches to traditional furnituremaking problems. Says Carpenter, "I think it is important to be an independent producer; I don't care what the product is as long as you're your own person. The more independent people there are about, the better off our society."

JAMES KRENOV

James Krenov serves as the country's most fervent voice for craftsmen who resist the economic and stylistic pressures of the marketplace to create their own deceptively simple, well-crafted furniture. Krenov calls these woodworkers "quiet" craftsmen and "impractical" cabinetmakers to distinguish them from those who emphasize innovation and expediency in their work. His ardent love of the material and his passionate defense of hand tools over more "efficient" machine methods have earned him a wide following both here and abroad, especially among the growing ranks of serious amateur craftsmen.

Although Krenov has communicated his ideas through years of teaching and lecturing, he is best known for the four books and many articles he has written, which articulate his philosophy on the craft and present his work in numerous illustrations. Amid the plethora of dry how-to books on furnituremaking, Krenov's deeply personal style strikes such a responsive chord among readers that hundreds have written to express their appreciation for the inspiration and support he has lent them. Regarding Krenov's writing, John Kelsey, former editor of *Fine Woodworking*

Magazine, once told him, "You have this gift . . .of conveying enthusiasm and insight and inspiration by the passion in your voice."

In addition to the practical information his books supply on methods, tools and wood, Krenov includes what he calls a "message of calm patience and of the exciting search that is the *raison d'être* for some craftsmen." Krenov believes most galleries and craft publications disregard fine craftsmanship in favor of "contrived designs and engineered woodworking." According to Krenov, "So much is said about design, derivation of style and the top-heavy technical emphasis on our craft, pushing some of us toward the ultimate efficiency: the Atari wood-worker To them I say, put the word 'intuition' into a computer and see what comes out."

Krenov identifies two divergent groups within the woodworking craft. In one group are the many ambitious woodworkers who rely on machine processes in their efforts to be "new and striking" and to make a name in the competitive environment of contemporary crafts; the other contains many amateur and semiprofessionals who are drawn to the craft by a fascination for the material. They express themselves through the way they work rather than through the forms they produce. Krenov has become the symbol of this latter group and his writing is dedicated to encouraging and guiding their efforts.

The highly opinionated Krenov does not hesitate to criticize the most famous figures in American woodworking. Of

Krenov has made seven silver chests since 1961, changing the configuration of the drawers and shape of the legs each time. The runners extend beyond the drawer fronts to form pulls. This chest is made of American spalted maple with bird's-eye-maple drawer fronts (1977).

This cabinet of American spalted maple (1982) contains many subtle curves and details, indicative of the personal touches Krenov puts into his furniture. The door panels, for example, curve out while the door frames curve in. The cabinet can be lifted off its red-oak stand, if necessary. The drawers are Lebanon cedar except for the fronts, which are partridge wood.

Sam Maloof he says, "anybody who makes 40 to 50 pieces a year is closer to manufacturing than to the best one-of-a-kind craftsmanship." He chides George Nakashima for failing to see "that there is more to woodworking than reverence for trees." However, he reserves his sharpest rebukes for Wendell Castle, who epitomizes Krenov's image of a craftsman who has abandoned the "integrity and humility of his craft to achieve commercial success by overplaying the 'art' of novel forms—some of which I consider more borrowed than new."

Krenov's furniture is an orchestration of details and subtleties that reveal themselves upon close examination. Examples include the intricate framework supporting his American-spalted-maple cabinet, the gentle taper of the legs of his writing desk and the thoughtful placement of a drawer, as in the Ballet Dancer cabinet shown on the next page that permits access even when only one door is open. These personal touches he says "are a joy to create and seem to be a joy for others to discover and rediscover."

The delicate door and drawer pulls that Krenov incorporates into his work exemplify the care he takes with detail. Some pulls are sculpted from the door itself, while others, like those of his silver chest, shown on p. 101, sprout from the sides of the drawer. In the American-spalted-maple cabinet he

The precise joinery of this jewelry box of Andaman padauk (1969) illustrates Krenov's dedication to craftsmanship. The hinges and hardware are silver.

JAMES KRENOV

This graceful cabinet of spalted maple (1978) is called the Ballet Dancer because of the long, slender legs that seem to be standing on tiptoe. Krenov chose secupira for the stand and handles because he felt its ''misty texture'' would complement the maple. The drawers are Lebanon cedar with Indian laurel fronts.

made in 1982, he salvaged the narrow stem of a lilac bush to shape two small pulls from its ivory-colored wood. Unlike most hardware, which detracts from a piece, the fixtures Krenov carves add character to the work and enhance its theme.

Opinions differ widely on Krenov's use of scale. The designer-craftsman Judy Kensley McKie sees ''elegance'' in his proportions. ''It's an intuitive thing,'' she explains, ''I don't think he takes his proportions from Greek forms like some people—he has a natural feel for

it.'' While some view his sense of scale as masterful, the words ''clumsy'' and ''ungainly'' have also been used to characterize his cabinets. Morris Sheppard, a versatile young furnituremaker who frequently employs a mathematical approach to design, says, ''there is an entire science of proportion that Krenov chooses to ignore, and you can see the lack of it in his work.''

Instead of working from precise shop drawings, most of Krenov's furniture evolves as he works. ''I am not much for drawing. . . . My pieces are

composed around an idea, a feeling—
and a chosen piece of wood. All the
little details, the way things add up, are
unpredictable—or nearly so. It is a
fingertip adventure."

Frequently, he repeats a basic design,
but changes it to suit the wood and his
mood. Commenting on his silver chest—
he has made about six of these pieces
over the last 30 years—Krenov wrote
". . . I do not use templates or jigs, but
work by eye. Each time I shape the legs,
tapering from the floor up, they are a
bit different. . . . "

If Krenov originally created a cabinet
in a light wood such as maple or pear
wood, he may remake it years later in a
darker one such as walnut or rosewood,
adjusting the dimensions to fit its new
character. He also uses contrasting
woods within a piece. "I am constantly
working with these elements: light and
dark shading. A tension in the wood
and then another tension to respond to
it, to play with it in a way that I feel
has some meaning."

One board often serves as the genesis
of an entire cabinet, as in the most

Krenov continues to work on an
increasingly larger scale. This
Japanese-oak cabinet with lignum-
vitae handles (1984) was a remake
of a cabinet he originally made in
1967 of lemonwood.

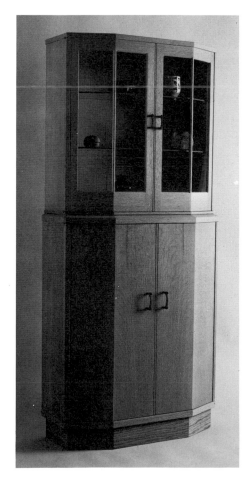

JAMES KRENOV

Krenov bandsawed his own veneer to about ³/₃₂-in. thick for this No Glass Showcase of lemon and doussie wood (1962). The cabinet contains two secret compartments, which can only be reached by removing the drawers and releasing a hidden latch.

recent version of his Ballet Dancer cabinet, made in 1978, with its fascinating spalted-maple doors. Says Krenov, "You get the impulse from the very texture and rhythm and color and feel of a particular wood, a particular plank, a particular piece of it." Wood's infinite variations are a constant source of awe to him, not only among species, but within one type of wood as well. "They vary from tree to tree, just as one person varies from another. . . this is the first joy of wood, essential to the patience and energy we need in order to work with it while listening to it."

Krenov's feeling for hand tools links him with the long history of his profession. He is especially fond of hand planes—which he makes himself—calling them "the cabinetmaker's violin; the instrument that sets the tone of the music in an orchestra."

In contrast to many craftsmen who take pride in their speed and production, Krenov has rarely made more than six to eight major pieces a year. "Where did we get the idea that fine things are made quickly?" he asks.

Krenov seems to gravitate toward pieces requiring time and patience, such as this chess table (1970). The frame is secupira and the playing surface is pearwood and Rio rosewood. Two maple drawers hold the chess pieces.

JAMES KRENOV

Krenov was born in Siberia in 1920, where his aristocratic parents were stranded during the outbreak of the Russian Revolution. When he was three, his family emigrated to Shanghai, China, and then to Seattle, Washington. Soon after, they settled in the Alaskan villages of Sleetmute and Tyonek, where his parents taught for the Bureau of Indian Affairs.

Despite their meager income during the lean years of the Depression, Krenov's mother retained her attach-ment for finely made objects. He fondly remembers her scrimping to buy two gold-brocade chairs she found in a second-hand shop. "To her, the genuine transcended regions and classes and was central in the appreciation of true qual-ity," says Krenov—a theme he reaffirms in his writing and teaching.

Around 1932, he and his mother moved back to Seattle where he grew up admiring ships and sailing. As a young man he worked in a shipyard on Lake Union building and refitting fine wooden yachts. It was his first serious experience working with wood and helped shape his sense of design. "[Ships and boats] had a definite and lasting effect upon me," he wrote, "with their grace or lack of grace, their symmetry—not that perfect symmetry, but always something alive. . . the straight line that was not quite straight; it was just enough strain by intent and by use to be alive, to have a message for whoever understood."

At the age of 27, Krenov left the

Krenov was commissioned to cre-ate this box of Swedish maple (c. 1964) to store the pottery col-lection of the late King Gustav of Sweden.

United States to see Europe and landed in Sweden. By performing piecework in an electric-light factory, he earned enough to travel throughout Sweden, France and Italy, and subsequently wrote a travel book on Italy, which was published in 1955.

Still undecided about a career, he stumbled upon the showroom of Carl Malmsten, a master craftsman whose name was then synonymous with contemporary Swedish furniture. Krenov later enrolled in Malmsten's cabinet-

Known as the Pagoda cabinet (1971), this cabinet of cherry with Lebanon-cedar drawers is among Krenov's favorite works.

making school. Although he received a solid background in traditional cabinetmaking, the school offered little encouragement for individual creativity. Following his graduation in 1955, he worked for Malmsten's son designing and constructing architectural models, and eventually established a small, one-man shop in Bromma, a suburb of Stockholm.

He chose for himself an area of furnituremaking—small cabinets and showcases—that he felt was ignored by most woodworkers and required a challenging variety of skills. Most of his work was made on speculation in hopes someone would appreciate the careful, loving way he constructed each piece. During those early, lean years, his wife, Britta—whom he married in 1951—returned to school to become a teacher to support them.

Although he gained many supporters and his work was included in collections of Sweden's Nationalmuseum, Sundsvalls Museum and Röhsska

In Krenov's unusual standing clock, the face appears to bend the willowy stand like a fishing pole.

Although known principally for his cabinets, Krenov occasionally attempts other works, such as this writing desk of Italian walnut (1977).

JAMES KRENOV

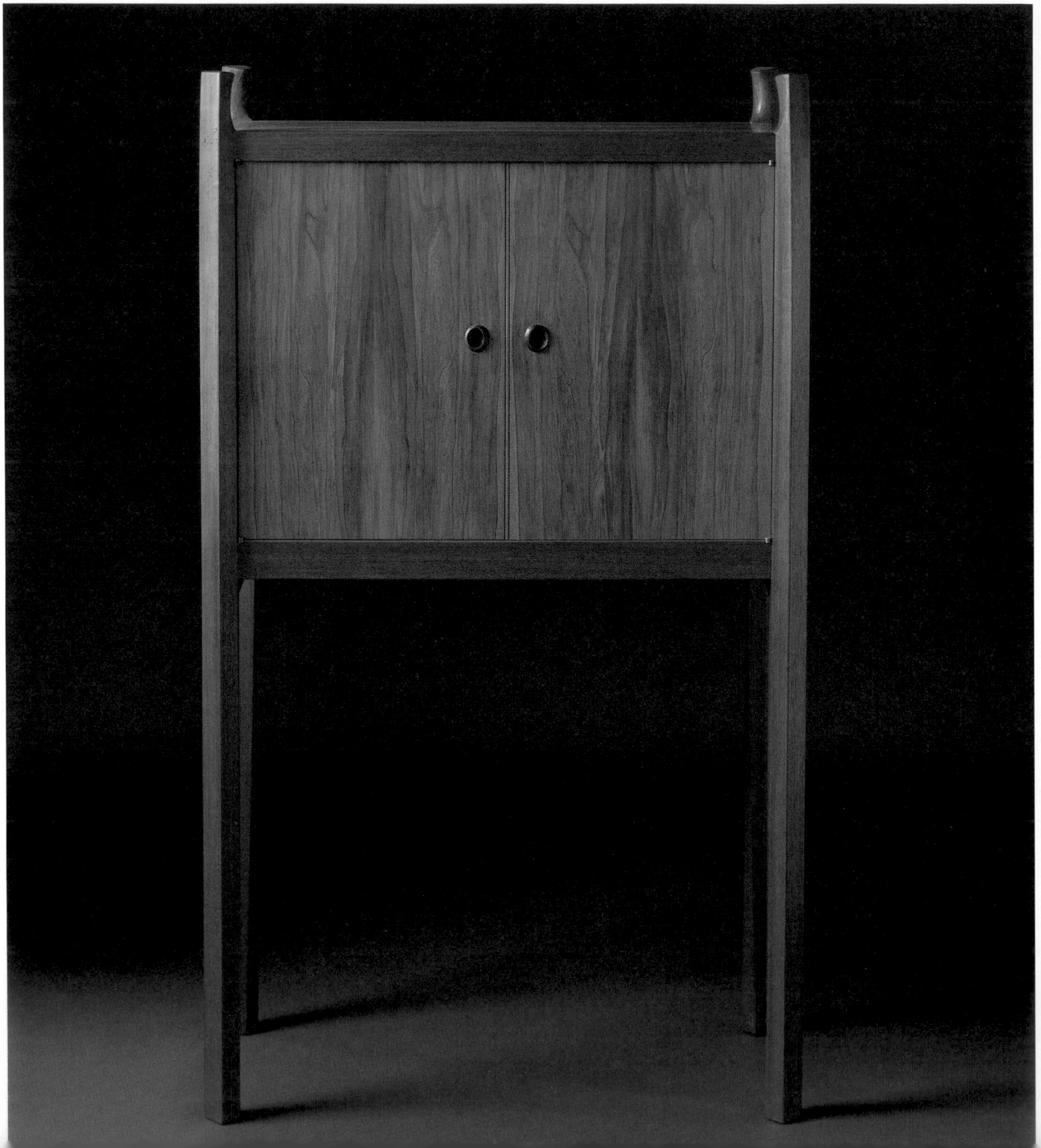

Konstslöjdmuseet, the straight-laced Swedish craft establishment never embraced his unorthodox ideals. *Swedish Interior,* a trade journal, said of Krenov, "Despite the chilly cultural climate in Sweden, James Krenov is admired and in many ways praised and appreciated. But at the same time, his really unique knowledge is shamefully neglected and the enthusiasm he retains and would happily share is ignored."

Craig McArt, professor of industrial design at Rochester Institute of Technol-

The veneer of this yaca-wood cabinet (1983) was bandsawn. The legs and frame are Honduras mahogany and the drawers are Lebanon cedar with Brazilian rosewood fronts.

ogy, met Krenov in 1965 while McArt was in Stockholm studying Scandinavian furniture production. McArt persuaded RIT to invite Krenov to America to teach at the School for American Craftsmen in 1968 while William Keyser, Jr. was on sabbatical. Krenov returned to teach there during the summers of 1973 and 1974.

With the encouragement of students and colleagues, Krenov recorded his ideas and experiences as a designer-craftsman for publication. (By this time, he had already written a few short stories and several articles on furniture for *Form* and *Craft Horizons* magazines.) *A Cabinetmaker's Notebook,* published in 1976, was the result. The response was overwhelming, and based on

Notebook's success, he wrote three additional books: *The Fine Art of Cabinetmaking* (1977), *The Impractical Cabinetmaker* (1979) and *James Krenov, Worker in Wood* (1981).

During the late 1970s and early 1980s, Krenov lectured and presented workshops throughout the United States and abroad, but maintained permanent residence in Sweden. In 1975, he helped create the wood section of the Program in Artisanry, then located at Boston University. From 1979 to 1981, he led summer workshops in Mendocino, California—a small coastal town in northern California that boasts a major craft community—sponsored by the Mendocino Woodworkers Association and the College of the Redwoods. The seminars evolved into a formal woodworking program within the College, and Krenov has served as its director since 1981. The Krenovs live in Fort Bragg, California. They have two daughters, Katya and Tina, and two grandchildren. Krenov shares a small shop near the College.

In direct conflict with the dominant trends of modern craft, which seek to "elevate" woodworking to an art form, Krenov advocates a return to the values traditionally associated with cabinetmaking, such as integrity, durability and independence. He personalizes each facet of his craft: Every board he selects, every joint he cuts, every detail he creates is performed in a way to immerse himself in the process of making. The result is a craftsman totally inseparable from his work and secure in knowing he has done his best and done it by himself.

WENDELL CASTLE

Wendell Castle is at the forefront of today's art-furniture movement, and more than any other contemporary woodworker he has redefined and expanded American attitudes toward art and furniture. His early, laminated pieces ignited the imaginations of woodworkers across the country and horrified many traditional craftsmen who saw their cherished techniques and conventional forms swept aside. Later, his *trompe l'oeil* pieces—still lifes of everyday objects made entirely from wood—established his reputation as an artist-furnituremaker and engendered his interest in historical furniture styles. His series of art-deco pieces attracted enormous attention with its revival of traditional forms, classic detailing and meticulous craftsmanship, all updated to reflect contemporary tastes. Experimentation with architectural elements led Castle to his most recent work, which continues to implement his philosophy that furniture can transcend its utilitarian role and communicate a message.

Despite the seemingly divergent styles of Castle's body of work, there is a common thread. Historically, craftsmen have felt that their role is to satisfy a practical need. But as an artist, as well as a craftsman, Castle believes his primary obligation is to fulfill an aesthetic need, and his furniture reflects this. This is not to say that his furniture does not meet minimum functional requirements—it does. For as Castle says, "I don't start with the function, I take that for granted. A chair has to be suitable and a table has to be able to support something or it wouldn't be a table. Now in some cases . . . they won't hold very much, but they will hold something." Castle's chairs are, in fact, more comfortable than those of many craftsmen. About this he says, "I feel the chairs I make are comfortable, even though I don't spend a lot of time thinking about that part of it. That part seems to come rather naturally. I spend more time thinking about how I'm going to suspend this thing in air to hold a person."

Like Wharton Esherick, Castle's first love was painting and sculpture, not furnituremaking. Castle had made a few pieces of wood furniture during his studies as an industrial designer as well as some wood sculpture as part of his fine-arts training, but it wasn't until he discovered Esherick's work in 1958 that Castle considered furniture as a form of sculpture. "Seeing Esherick's work made me realize furniture could be a means of self-expression. Until then, I never thought it possible to make a living creating art furniture." Although only a sideline at the time, it was Castle's early furniture, not his sculpture, that generated sales. So, like Esherick, Castle concentrated on incorporating his sculpture into his furniture.

Castle's early furniture relied heavily on organic shapes. His music stand (see p. 124), was his first piece to

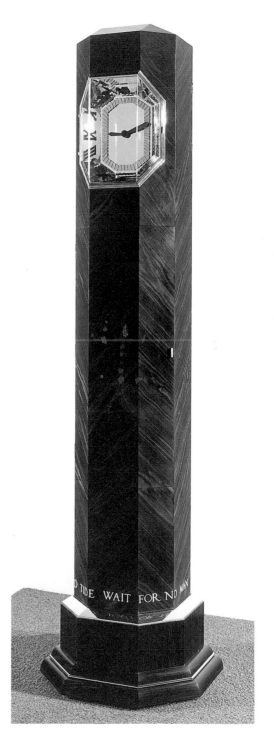

This three-seat settee (1968) illustrates Castle's interest in organic forms. The entire piece is stack-laminated cherry.

The first and most traditional of Castle's long-case clock series was this Octagon clock (1984) made of Indian ebony, capomo, sterling silver, brass and gold and silver plate.

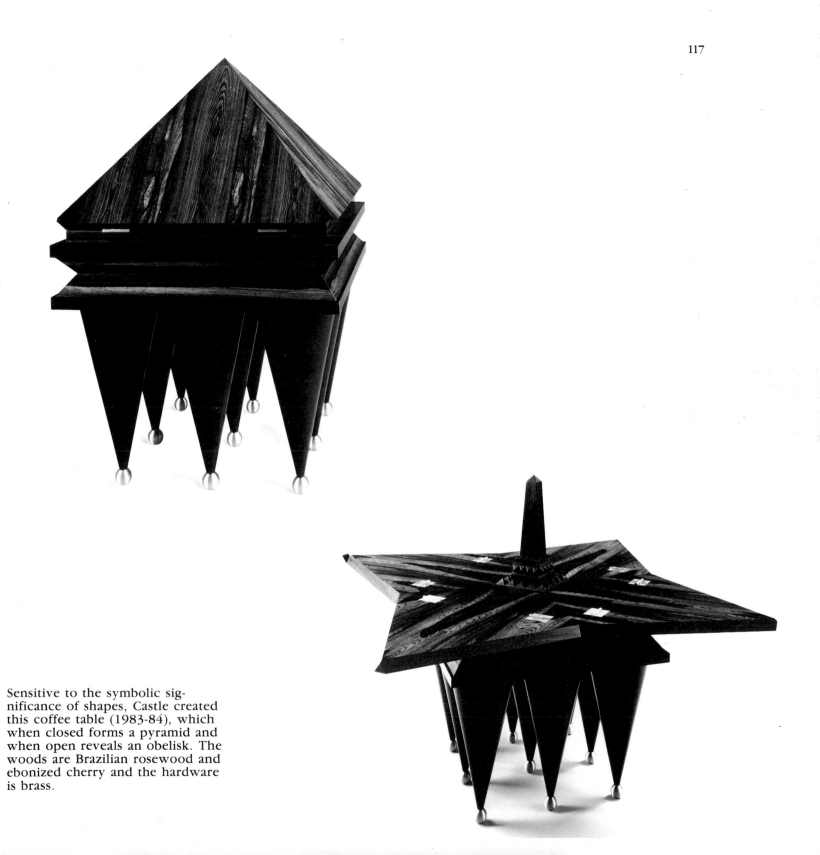

Sensitive to the symbolic significance of shapes, Castle created this coffee table (1983-84), which when closed forms a pyramid and when open reveals an obelisk. The woods are Brazilian rosewood and ebonized cherry and the hardware is brass.

This maple dining table (1978) was built one layer at a time. After one lamination was bandsawn to shape and glued to the previous layer, it was resurfaced to receive the next lamination. The rough form was shaped with a power plane and Surform and then finished by electric and hand sanding.

achieve wide recognition; it was also the subject of an award-winning educational film by Tom Muir Wilson in 1964. "As I originally conceived the idea," Castle remembers, "the music rack would be an artistic investigation of a treelike form. The legs would be rootlike, embracing the ground in a gentle, but regular, organic curve. From them the trunk would ascend like a supple, bent sapling. A twiglike appendage would descend to hold the easel."

The technique of bent lamination allowed Castle to create the music stand's graceful curves—Castle glued together ⅛-inch-thick strips of wood, clamped them around a mold, then used a spokeshave and rasp for final-shaping. To Castle, lamination was a

perfect way to overcome the structural limitations of wood, and thus was the key to creating sculptural forms. In his book, *The Wendell Castle Book of Wood Lamination*, which he co-authored with David Edman, he discusses the advantages of lamination: "It was a combination of these qualities—strength, stability, mass—that began to . . . appeal to me in the early 1960s. Strength and stability are always welcome additions to woodworking procedures, for wood in its natural state is only fitfully strong and is notoriously unstable. As for an increase in wood mass, it seemed to me that here lie possibilities for entirely new approaches to wood as a medium of artistic expression."

Castle achieved the mass needed for his large works by dividing the form into many layers, each carefully planned and cut so when stacked atop each other they formed the basic shape. This technique, which Castle called stack lamination, originated from the model World War II airplanes he built as a boy from balsa wood. Soon woodworkers across the country were trying stack lamination in their workshops, whereas the technique had been nearly unknown in furniture just a few years earlier.

Defenders of traditional woodworking methods, such as Tage Frid, cautioned against the technology and predicted the pieces would eventually fall apart. But after 25 years, Castle reports practically no problems with his stacked pieces. He attributes this to the elaborate precautions he took in preparing the stock—to achieve the strongest glue joints, the wood's moisture content was meticulously monitored and the wood was glued up immediately after planing.

The designs for Castle's laminated pieces were drawn from flowers, trees, bones and other natural forms. The legs of his enormous three-seat settee, (see p. 114), appear to be reaching out like roots in search of water. (The cupped shape of the seats would have been almost impossible to create with traditional woodworking joinery.) And a whelk shell inspired his maple dining table. Like many of Castle's tables and Castle feels the Victory desk (1978), his last major stacked piece, fulfills the artistic goals he sets for his furniture. He named the desk after the classic sculpture Winged Victory, because he felt the desk exhibited the same flowing lines as the statue.

WENDELL CASTLE

Like the popular Crescent rocker, this Zephyr rocker was produced in signed editions by Castle's workshop in Rochester, New York. This chair (1974) is curly maple.

Six pie-shaped pieces come together at one point on the top and bottom of Castle's Brazilian-rosewood *demilune* table (1982). The elegant form and use of ivory detailing are reminiscent of the art-deco designs of the 1920s.

desks, the top is simple compared to the intricate base. But despite its illusion of weight and mass, the base is entirely hollow, created from a stack of many layers carefully precut to rough shape. A series of final-shaping processes, which began with a chain saw, pneumatic chisel and body grinder, and ended with various hand tools and a week of hand-sanding, refined the form. In the Victory desk, Castle's last major stacked piece, built in 1978, the graceful, flowing form supersedes all other considerations—the flat top seems almost an afterthought.

The Alpha chair (p. 126), introduced in 1964, was the first chair design Castle repeated. With its conventional four legs, it's the most conservative piece of his stacked period. But despite its traditional elements, it follows the same organic course as Castle's other work of that time—the sides fuse together in a wishbone configuration and the seat gracefully tapers into the uprights.

Introduced in 1972, Castle's Crescent rocker has become his most popular design. Although he has produced more of these rockers than any other design in his repertoire, he considers the rocker among his least significant work. "It doesn't push back the limits of design," says Castle. "To me, it's too much like furniture."

Castle's stacked pieces reflected their time. The 1960s and early 1970s saw not only social rebellion but fashionable discontent in the art world as well. "It was the last vestige of the artist as a starving individualist expecting everyone else to come around to his thinking," says Castle. "Art didn't have to

communicate—art was art. I was part of that ideology. Today, art is more like a business and now I use decorative motifs to convey my message rather than the idea that art represents art." Eventually, however, Castle felt he had taken stack lamination as far as he could. "By the mid-seventies, I became disillusioned with it. There was no challenge to it any more—the process became repetitive. The germ of my current direction crept into my work even before I was aware of it." His graceful Pinwheel game table and three-legged writing table, designed while he was still involved with stack lamination, foretold an interest in more classical furniture forms.

In searching for a new direction, Castle conceived his *trompe l'oeil* (fool the eye) wooden still lifes, among them a crumpled fedora resting on a Chippendale chair and a sportscoat draped over a schoolboy chair. The pieces combined stack lamination with conventional woodworking methods, and the carvings were so realistic, they often fooled visitors to Castle's workshop even before they were completed.

The series generated national publicity. For 20 years, Castle had sought the approval of the art world with his sculptural furniture without success, but ironically it wasn't until his work appeared more like traditional furniture that it was finally accepted as art. "I had thought the way to have my work recognized as art was to make my furniture look like sculpture, but it wasn't until my sculpture looked like furniture that it was finally accepted as art."

In an unexpected way, the *trompe*

WENDELL CASTLE

When creating his *trompe l'oeil* still lifes, Castle included readily recognizable elements such as Chippendale chairs and a squashed hat (1980) carved entirely from Swiss pearwood.

Castle was particularly fond of carving coats in his *trompe l'oeil* still lifes, such as this man's suit jacket slung on the corner of an unpretentious looking chair (c.1979) in maple.

l'oeil pieces exposed Castle to classic furniture, which in turn engendered an entirely new direction in his work. To create the illusion he wanted, he created the still lifes from readily recognizable objects. "I felt the furniture should have a historical reference. People perceive things as they *think* they are—not necessarily the way they really are. If a piece looked more or less like a Sheraton, it would be perceived as one. But I didn't know enough about furniture styles to do that, so I did some research. This opened up a new learning cycle for me; until then my art had fed itself." This represented a total reversal in attitude—whereas once Castle had rejected historical styles as a source for design ideas, he now embraced them.

Of the many periods he studied, art deco struck the most responsive chord, particularly the work of Jacques-Emile Ruhlmann, the last of the great French *ébénistes.* Ruhlmann had revived the seventeenth-century French cabinet-maker's passion for detailing and meticulous craftsmanship, especially in veneering. But unlike most of his predecessors, Ruhlmann used detailing to accentuate a piece's form rather than simply to embellish it.

Castle's first work to make use of detailing in this fashion was the lady's writing desk and accompanying chairs made in 1981. Although rows of dots and slippered feet are decorative devices appearing in fine furniture for centuries, in Castle's hands they exude a decidedly contemporary flavor. "I'm continuing the furniture tradition using the same decorative vocabulary that's

been around for 400 years but keeping it in step with the times," he says.

Like Ruhlmann, Castle has succeeded in changing the perception that the only valuable furniture is fine antiques—the writing desk and chairs carried a $75,000 price tag, and other

pieces of Castle's work command prices up to $100,000 or more. The cost reflects the materials and enormous overhead of his shop, including a staff of 10 craftsmen. One piece may require 500 to 1,000 hours of labor. Does he feel uneasy that his work is beyond the reach of most people? "In Ruhlmann's day, the price of one of his pieces was about that of a house in Paris," he replies. "I'm making the best furniture I can, using the finest materials and employing skilled craftsmen. To sell it for less I'd have to cheapen the product."

Castle's investigations of historic furniture styles for his *trompe l'oeil* series inspired him to new directions. This elaborate Lady's writing desk and accompanying chairs (1981) incorporate traditional feet and decorative elements such as 8,500 ebony dots. The wood is curly English sycamore and the drawers are amaranth. The chairs are upholstered in Jack Lenore Larsen brocade.

WENDELL CASTLE

Castle calls this clock the Magician's Birthday (1984) because it appears as though the numbers were whisked off the face of the clock and magically transmuted around its 12 feet. The woods are ebonized cherry, East Indian rosewood and Gabon ebony with a gold-plated brass movement.

During 1963, when Castle's stacked period was in full swing, he also designed this light, simple music stand. It is composed of just eight pieces: two legs holding a main stem in oak and a branch supporting vertical rosewood slats that hold the music. Every part is laminated and clamped around a mold to create the curves.

A variation of the Pinwheel game table, this walnut and satinwood writing desk with silver-inlay detail (1984) is a modern adaptation of a conventional furniture form.

Among the most notable work Castle produced in 1982 were two sensitively proportioned and detailed designs: the *demilune* table (p. 121) and the Fountain cabinet (p. x). The ivory inlay and fixtures at the tope of the legs of the *demilune* table are especially reminiscent of Ruhlmann. But the dramatic Fountain cabinet, wityh its radiating pattern of sterling-silver inlay on an English walnut case, is another example of a contemporary update of a classic deco design. (Another version of the Fountain cabinet has ebony dots on a pearwood case.) Depending on the distance the viewer stands from the work, the dots appear to merge into solid lines, suggesting Castle's fascination with perspective and illusion. (The pattern is repeated on the back of the pearwood cabinet, producing the same

effect from behind when the doors are open.)

In 1982, Castle expanded his design vocabulary to include architectural elements. With its heavy, tapering column and sturdy base surrounded by black rings resembling machine bolts on end, his Octagon table utterly abandons the graceful fragility of art deco. Only the inlaid ebony dots recall the delicate designs he created just months earlier.

Castle continued to explore geometry with an impressive liquor cabinet that rests on four inverted pyramids (p. 129) and a jewelry box titled Late Proposal for the Rochester Convention Center. The box is a whimsical comment on architecture. Following the opening of Rochester's new convention center, Castle felt that he could design a

This bird's-eye-maple and ebony Octagon table (1982) is in the permanent collection of the Lannam Foundation in Florida.

WENDELL CASTLE

Castle's Alpha dining chair of ebonized oak (1979) exhibits many of the same qualities of his stacked works: The seat flows into the legs, which in turn spread and meet the ground like a newborn colt gaining its balance.

Castle titled this lacquered jewelry box Late Proposal for the Rochester Convention Center (1982). The wood is Carpathian elm with gold-plated brass hardware.

better one. The center was located at the junction of a major street and a river. Castle designed his convention center so it rested on four bagel-shaped points spanning both the river and the road, and painted it in Howard-Johnson blue and orange. Says Castle, "If they had built my design, it would have put Rochester on the map."

Similarly, his folding coffee table (p. 117) incorporates many architectural elements. When closed, it forms a pyramid; when open, it reveals an obelisk—two shapes, Castle points out, rich in mystical connotations. "The obelisk is a monument," he says. "It can commemorate a victory or a death." The base of the coffee table is composed of nine cones resting on brass balls. The use of metal for decorative detailing occurs with greater frequency in Castle's most recent work. (Garry Knox Bennett, who shares Castle's approach to furnituremaking, has assisted Castle with metalwork.)

In the tradition of the great cabinet-makers, Castle steadfastly refuses to compromise his high standards of quality, but fine craftsmanship is not enough for Castle. "Craftsmanship is a means, not an end," he says. As an artist, he feels compelled to continuously break new ground. His latest work, a series of long-case clocks, attempts to merge his two recent interests: the eloquence of art deco and the boldness of architecture.

To Castle, clocks are poetic images representing the passage of time and how people spend their lives. His series consists of 13 variations of long-case, or grandfather, clocks, which Castle uses as surrogates for the human figure. "Long-case clocks are a way to present the human figure, which, according to the classic and poetic ideal, is the perfect image. They have a face, hands and a body of roughly the same proportions as the human body. Some even have feet." The clocks were designed to be

shown as a group. All employ exotic materials, including precious metals, and involve geometric forms, especially pyramids and triangles. Castle commissioned a poem for each clock by the eminent British art historian Edward Lucie-Smith.

The series begins with the Octagon clock (see p. 115), the most traditional of the group. Its powerful octagonal form acts like an anchor, linking the series with its heritage. From there, the forms become increasingly more exotic and occasionally playful. The Magician's Birthday clock (p. 124) is among the wildest, with four separate sets of hands driven by a complex custom-built mechanism. The legs and proportions of the clock are based on the number 12. A numbered bracelet encircles each leg as in the finale of a magic trick. The series ends with The Ghost, a *trompe l'oeil* image of a grandfather clock hidden beneath a sheet. "Clocks often contain spirits," says Castle.

All 13 clocks were designed and constructed within one year, illustrating the remarkable speed with which Castle conceives and executes his ideas. New designs begin as rough sketches or models, which he converts into full-size, but still rough, drawings. Shop assistants translate the sketches into precise working drawings. Castle reviews the work at each step to make sure it follows his plan, and the final piece is remarkably close to the original rough sketch or model.

Castle was born in Emporia, Kansas, in 1932 and studied industrial design at the University of Kansas. "I would have preferred to study painting but my family wanted me to pursue something employable. Industrial design seemed a logical compromise." After graduating in 1958, he worked for an aerospace firm on a project to design a moon base. "I realized at that point I really wanted to be involved in making things, not just designing them." He returned to the

University of Kansas to study sculpture in 1961 and earned a master of fine arts degree. Immediately after graduation, he moved to New York City, "where all big artists go."

Sobered by his failure to make a big splash in the New York art world, he secured a teaching position at the School for American Craftsmen of the Rochester Institute of Technology in 1962, replacing Tage Frid in the wood department. He had been making furniture on a part-time basis for four years. In 1969, he left RIT for a position with the State University of New York at Brockport. By this time he had his own shop employing three people and was producing about 75 pieces a year.

Unlike Sam Maloof, who refuses to relinquish any part of the technical process, believing the best results occur when the designer is also the maker, Castle feels his time is best spent on more creative activities. So Castle has always hired other craftsmen to help

The dentil-like pattern on Castle's monumental Egyptian desk (1982) continues across the front of its three drawers. There are also two secret compartments, but the owner won't say where. The desk is made of Macassar ebony, maple, ebonized maple trim and gold-leafed mahogany medallions and leather top. The entire desk easily disassembles into five pieces for transport.

WENDELL CASTLE

him, although initially they served more as shop assistants than cabinetmakers. During his stack-lamination period, he developed the designs and performed most of the shaping, while his assistants would do the milling and sanding. Today he employs 10 accomplished cabinetmakers who assume nearly all the technical work under his supervision, leaving him free to concentrate on design and management. Don Sottile, who has been with Castle since 1976, assists him in running the operation. Dividing the work among several craftsmen expands the capability of the shop and increases its output. "The whole idea of the craftsman has become very narrow. In the seventeenth century, the cabinetmaking guild was completely divided. It was never expected that the same person would do all the joinery or veneering or carving. It wasn't permitted. We've done something like that on a limited scale in this shop, making use of people's individual talents." Castle himself, however, delves into every technical area.

Throughout his career, Castle had often considered establishing a school that offered a curriculum balanced between technical and design courses, which he felt was lacking in most craft programs. "Most craft programs foster a self-centered attitude among their students where they feel they must do everything. As a result, they can't work as a team. Everyone shouldn't be expected to be good at everything. If a student isn't a designer, don't encourage him to be one. If he isn't a craftsman, don't encourage him to be that." He resigned his university position in 1980

and opened the Wendell Castle Workshop that fall in Scottsville, New York. Most of the program's 30 full-time students enroll for two years of highly structured curriculum, although a third year of independent study is possible. The faculty consists of three full-time and two part-time instructors. Castle spends about a third of his time teaching, with the balance split equally between administration and working in his studio.

The school is located in the same building as Castle's shop so the students learn in the atmosphere of a viable commercial operation. Every Friday afternoon, they can tour the workshop and ask questions of the craftsmen. A few of the more promising graduates receive offers to join Castle's staff. The school and studio have an impressive array of equipment, including industrial-size tablesaws, veneer presses, shapers, sanders, jointers, bandsaws, mortisers and planers, and a metal-milling machine—the shop produces much of its own hardware, such as hinges, latches and clock gears.

The huge, meandering structure that houses the school and studio was once a bean mill and dates back to the 1890s. Until 1979, Castle and his wife, potter Nancy Jurs, and their daughter, Alison, lived above the studio. They have since moved to a stately home nearby, which contains many examples of Castle's work along with samples of his wife's sculpture.

Castle has exhibited extensively and his work appears in the collections of the Metropolitan Museum of Art, the Museum of Modern Art, the

Smithsonian Institution, the American Craft Museum, Boston's Museum of Fine Arts, the Philadelphia Museum of Art, the Addison Gallery of American Art, the Chicago Art Institute, the Houston Museum of Fine Arts, Nordenfieldske Kunstindustrimuseum in Norway and the Patrick Lannam Foundation, among others. His awards include an honorary doctorate degree from the Maryland Institute of Art, the Lillian Fairchild Award from the University of Rochester and grants from the National Endowment for the Arts and the Louis Comfort Tiffany Foundation.

Castle is critical of the furniture produced today. "I see little that stands up to 200- and 300-year-old pieces," he says. He aspires to provide the innovative leadership he believes is lacking in today's furniture craft. "I'm not trying to be the best furnituremaker on the block—I prefer to make a design contribution. Going back to the late seventeenth century, the great cabinetmakers set the standard and tone, which hundreds would then follow. That's what puts work on the cutting edge."

In the early 1980s, Castle became interested in geometric forms, especially pyramids, exemplified by this liquor cabinet (1982) of Australian lacewood, amaranth, gold leaf and gold-plated brass hardware.

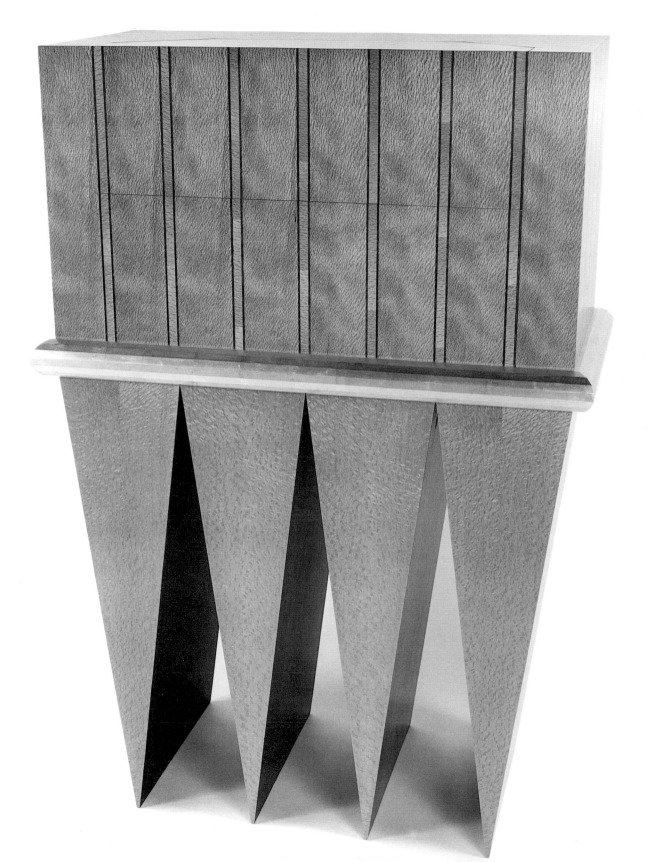

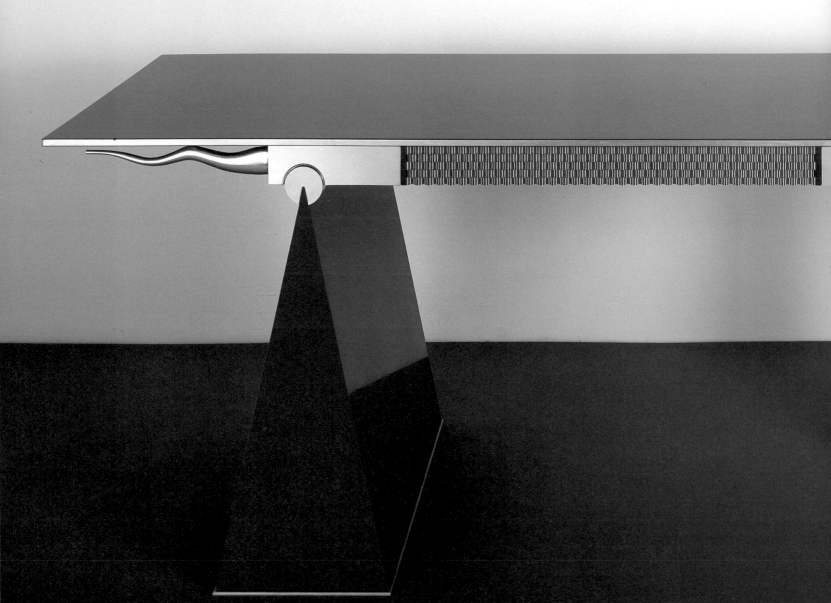

GARRY KNOX BENNETT

Garry Bennett breaks all the rules. Because he came to the craft via a convoluted route of sculpture and metalworking, he holds no preconceived notions of how furniture is made or what it should look like. His use of form, color, texture and materials is freer than that of most American woodworkers because he is not limited by sentimentality for the material or a formal education in its methods. His irreverence for traditional woodworking borders on hostility. Although he works predominately in wood, he prefers to think of himself as a furnituremaker rather than a woodworker. To him, wood is merely a "convenient medium for making large objects."

Bennett's designs are a panoply of vibrant colors and bold shapes. He creates them piece by piece, which accounts for their feeling of spontaneity. Unlike most furnituremakers who develop their ideas with detailed shop drawings or precise scale models, Bennett sketches directly on the wood. Wharton Esherick did this to take

Created for a 1984 exhibition at the Workbench Gallery in New York sponsored by the Formica Corp., this desk consists of an aluminum top covered with Colorcore® supported by a wedge-shaped upright of Colorcore® on particleboard at one end and a polished aluminum tube at the other. The rosewood drawers were constructed with dado joints, epoxy and bronze ring nails. The intricate pattern on the front is composed of multiple laminations of Colorcore® . The penholder and bar above the upright are 23k gold-plated brass.

GARRY KNOX BENNETT

advantage of the wood's color and grain, but Bennett does it out of an impatience to create. "I put pieces together in my head—what a table leg should look like or where a drawer should go," he says. "I work most of my designs out at night instead of counting sheep. Then I come into the shop the next day and start sawing." Yet Bennett never allows his impulsiveness to run wild—he maintains complete control, balancing contrasting elements so they accentuate each other.

Despite his background as a sculptor, Bennett describes his furniture as two-dimensional. "Generally, when I design a piece I'm concerned with how it looks from the front. If it's a bench, I don't even consider how it looks end-on because I assume it will go against a wall." Unlike his friend Art Carpenter, who advises woodworkers to turn their furniture upside down to analyze it from every angle, Bennett conceives his designs as line drawings. "It would be hard to eat on a table that's upside down," he replies dryly.

Bennett was born in 1934 in Alameda, California. His interest in art began in childhood. "I remember we had a substitute teacher in grammar school. She saw my penmanship and said, 'These are such beautiful curves. You should be an artist.' I couldn't have

Intended as a comment on craftsmen who have become carried away with themselves, Bennett's controversial Nail cabinet (1979) elicited the outraged reaction he sought. The case is padauk with a curved glass front and shelves and a lighted panel in back.

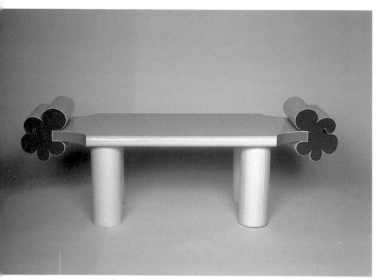

Because he used relatively soft woods—redwood and fir—Bennett painted the Mackintosh bench (1981) with automobile lacquers to give it a more resilient finish.

One of two benches Bennett designed honoring the contemporary sculptor Isamu Noguchi, Clouds on Noguchi (1981) is made of California walnut, Honduras mahogany and Honduras rosewood.

Bennett enjoys the challenge of creating wildly different designs based on the same form. The Checkerboard bench (1982) was one of a series of benches using the same templates.

Although the same basic shape a the Checkerboard bench, the Ri bon bench (1982) has an entirel different character. The wood i Honduras mahogany, poplar and walnut.

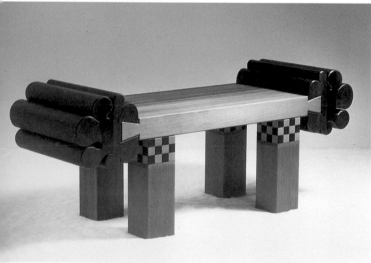

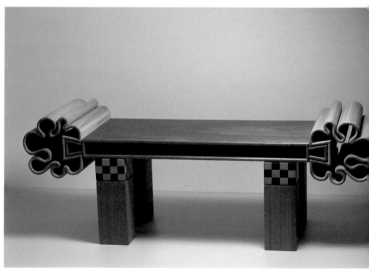

been more than seven or eight. She was pulling my leg, of course, but from that point on I wanted to be an artist."

He claims his high-school days revolved around cars and drinking and says he had an F average. Following his graduation, he worked the swing shift at the Foster Chemical Corporation. After a year of stacking 74-pound crates of kerosene, he decided college was an easier alternative and enrolled in the California College of Arts and Crafts from 1959 to 1962. While in college he married Sylvia Mangum, whom he had known since high school.

Although Bennett majored in metal sculpture, college afforded him a broad education in art and design. He recalls with fond disdain having to sketch the same plaster bust over and over again for an entire semester. "Maybe that's why I don't like to draw now, it's a reaction against all those academic art courses I had in school." But he admits his formal art training enhanced his ability to visualize and refine images in his mind.

He quit college four months before graduation—it was unfashionable to graduate during the rebellious 1960s—and moved to rural Lincoln, California, to become an artist and "live off the land." For eight months a year, during planting and harvesting, he worked on

Uninhibited by conventional furniture forms, Bennett created this cabinet with clock (1977-78) in walnut, cherry, and etched, galvanized steel.

GARRY KNOX BENNETT

his stepfather's rice farm for about 75 cents an hour. At night and during the off-season, he made sculpture. Eating little more than polished rice, he contracted beriberi from malnutrition.

Bennett's career changed direction in 1963, when he was introduced to marijuana. Inspired by its burgeoning popularity, he applied his metalworking skills to creating hand-crafted roach clips (marijuana-cigarette holders) and later broadened into jewelry. Based on the success of his metalwork, he borrowed $500 and moved back to Oakland in 1966 to establish Squirkenworks, a small electroplating company that mass-produced, among other objects, the first peace symbols as jewelry.

Having achieved financial success, Bennett searched for a more creative challenge and invented a graphic-art process he termed electroprinting. The technique combined his newly acquired knowledge of electroplating with his artistic skills. It involved using various chemicals and resists to color gold and silver plate. He later applied the process to small metal objects, notably clocks and kerosene lamps.

The clocks and lamps symbolized Bennett's transition from art to furniture. They were small enough to test new design ideas and could be made fast enough to hold his interest. "In a way, it was my most rewarding period because I could make one in a day or two. I get bored pretty quick," he con-

"I want my pieces to take over the room," is how Bennett summarizes his design philosophy. This conference table is Honduras mahogany and shedua. The glass top is 54 in. in diameter.

fesses. Like his furniture, the clocks consist of simple stylized shapes—almost cartoonlike—in bright, conspicuous colors. Many, like the clock with Terminal Susskind's Syndrome—the image of a bleeding heart—and the Bicentennial clock, were based on a literal theme. Often the concept was overtly sexual. There is nothing subtle about his Sperm Count clock or the Inter-Racial Cum clock.

Light fixtures and small chests of drawers represented Bennett's first experience with wood, which he adopted for entirely practical reasons. "There's no spiritual stuff involved," he maintains. "Wood is not as forgiving as metal, but it's a rapid medium for making fairly large objects." Yet he has

Although Bennett works rapidly, eager to move on to the next piece, these redwood and polychromed-pine doors (1976) captured his imagination and he spent many hours creating the pewter and bead detail.

GARRY KNOX BENNETT

Bennett combined his skill as a metal sculptor with his new interest in woodworking to create this standing light fixture (1979-80).

never abandoned metal, and often uses it interchangeably with wood—an anathema to purists like James Krenov and Sam Maloof, who avoid even metal hardware.

Metal dominated his standing light fixtures, his first large pieces of furniture, where an array of tubing sprouts from the sides of an otherwise unremarkable wooden cabinet. The joinery is primitive, relying mostly on nails.

A far more successful fusion of the two media occurs in a small chest of drawers (p.135). Its pyramidal top and triangular legs foretold his inclination toward unconventional furniture forms. He faced the top and drawers with galvanized steel etched in a delicate pattern of tiny brushstrokes. Here, metal has been elevated to a purely decorative role from its usually structural function in furniture.

Although these early furniture pieces were poorly constructed, they provided an opportunity to explore new forms and materials. As his work grew increasingly complex, he became more concerned with technique. "At first I used redwood and other humble woods and just threw it together. Then I felt guilty and started working slowly. I thought everything had to be made with *pain stakes*. I wasted hours." Now he looks for the most expedient way to create his pieces without compromising their

The rich wenge seat of this bench (1984) contrasts with the polychromed Douglas-fir uprights and drawer front. The interior of the yellow-satinwood drawer is polychromed with several layers of gesso and paint.

One of a series of trestle desk/tables
created by Bennett.

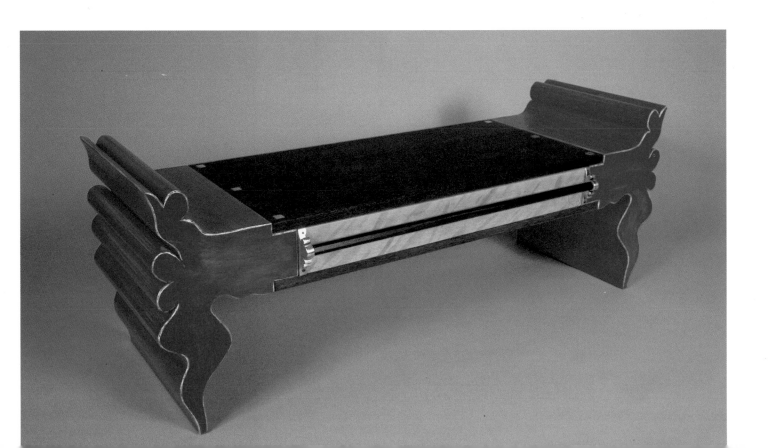

durability. "There's no labor of love here," he says, " 'cause I don't love labor." Rather than follow the traditional cabinetmaker's credo of "form follows construction," preached by Tage Frid and other East Coast university instructors, Bennett first conceives a form and then figures a way to make it. "I love the product," he says, "not the process."

In 1979, Bennett rebelled against what he saw as a growing emphasis on technique among craftsmen by pounding a 16-penny common nail into the front of a six-foot-tall, solid-padauk display cabinet that he built. With its finely dovetailed carcase, graceful curved front and secret latches, the case might otherwise have earned the admiration of cabinetmakers everywhere. "I wanted to make a statement that I thought people were getting a little too goddamn precious with their technique," he says. "I think tricky joinery is just to show, in most instances, you can do tricky joinery." To complete the irony, Bennett constructed the cabinet using several intricate woodworking methods including compound miters and curved laminations. When the controversial cabinet appeared on the back cover of *Fine Woodworking* magazine, dozens of outraged woodworkers wrote angry letters denouncing the "desecration." In an otherwise favorable review, *Artweek* magazine called it "either an apparent breach in the esthetic principles of its maker or a stab at the sensibilities of its viewer" (see p. 133).

In many ways, Bennett represents the antithesis of fellow Californian Sam Maloof. Besides their obvious differences in style and technique, they practice fundamentally opposing philosophies as well. Maloof prefers to refine and perfect his designs while Bennett constantly wants to move on to new ones. Yet both furnituremakers explore variations of a basic design. For Maloof, the changes are subtle; for Bennett, the variations are dramatic.

Benches were the first pieces Bennett made in series. They appealed to him because, like the clocks and lamps, he could make them quickly. In some instances, the form is the same but the surfaces differ widely. His Ribbon and Checkerboard benches—both made in 1982—illustrate different treatments of the same shape (p.134). Positive versus negative space is important in the Ribbon beach while decoration and texture dominate the Checkerboard one. He creates the checkerboard pattern simply by scoring the wood with a knife and

dying alternate squares—the technique is another assault on textbook woodworking, which would mandate inlaying contrasting pieces of wood to create the same effect. He combined plywood and gold leaf for a third bench with the same shape, as shown on p. 143.

Most benches, however, are one-of-a-kind, such as his Red bench (p. 139) which demonstrates his acumen with color. The fiery uprights constrast with the rich grain of the wenge seat and the

Bennett freely mixes wood with metal as in this desk with square legs of aluminum tubing and copper inlay and bubinga, ebony and Honduras mahogany. A dovetail projecting from the side of the drawer also forms the runner.

By adding a drawer to a trestle table, Bennett converts it into a desk.

GARRY KNOX BENNETT

elegant ebony and silver-plated drawer pulls.

Clouds on Noguchi and the Mackintosh bench (p. 134) parody the works of other designers. Clouds is one of two benches by Bennett honoring American sculptor-designer Isamu Noguchi. Like Bennett, Noguchi is concerned with mass and line so his forms easily adapt to Bennett's furniture. The Mackintosh bench satirizes Scottish furniture designer Charles Rennie Mackintosh, who used furniture to manipulate space and proportions.

The scale of Bennett's work continued to grow with a series of trestle tables with 7-foot-long by 3-foot-wide tops. Once again, he could make them quickly—about one a week—and still vary the design. He found the trestle a sound means of support that is interesting to look at and easily disassembled for transport. He takes great pleasure in creating elaborate wedge-and-tenon arrangements.

By adding a drawer and glass top, he converts tables into desks (see pp. 139, 141). "I love a glass top," he told *Fine Woodworking*. "It's thin and it lets you see what's underneath, where all that good woodworking got done." Bennett created a series of desks consisting of two triangular uprights sup-

Applying woodworking techniques to an entirely non-wood object, this brass, copper, and silver-plated table lamp (1985) stands 24 in. high.

porting a glass top and connected by a trestle. Here again, he varied everything except the basic form. The parts may be metal or wood, painted or natural, smooth or textured—depending on his mood. He is a man of infinite ideas and a short attention span. Even as he works on one piece, his mind races to the next.

Bennett's concern for form and line, color and texture, culminated in a desk he made with a delicately curved leg at one end and a massive black slab at the other (shown on p. xvi). The piece is a study in contrasts. The graceful taper of the leg balances the primitive quality of the slab; the freeform drawer fronts interplay with the desk's otherwise severe geometry; and the semi-spherical copper lamp offsets the warm walnut top. The black slab is California Claro walnut ebonized with black shoe dye—it still has the marks left by the lumbermill's saw, giving it additional texture.

"Bennett handles opposites brilliantly," says Wendell Castle, "positive and negative, rough and smooth, dull and shiny—playing them against each other." How does Bennett succeed in juxtaposing different elements without letting them appear disjointed? Castle believes it works because Bennett knows no fear and adds, "he's one of the loosest people in the field."

Bennett lives in Alameda, California, just a few blocks from where he was raised, and maintains a well-equipped shop in nearby Oakland. He signs his work "GKB—in Oakland anno . . ." proclaiming his loyalty to his hometown. His three-story Victorian house is crammed with contemporary paintings, sculpture and crafts, including woodwork by Edward Zucca, Bob Stocksdale, Wendell Castle, David Ellsworth and Judy Kensley McKie, among others. Bennett renovated parts of the house himself and added a redwood sun room that contains over 500 panels of glass. The Bennetts have three children, Aaron, Joshua and Jessica.

Bennett's work has been shown at numerous art and craft galleries and the Oakland Museum, the Craft and Folk Art Museum of Los Angeles, the Richmond Art Center, the San Francisco Museum of Modern Art and the American Craft Museum, among others. He is represented in the Oakland Museum and Oakland's Judah Magnus Museum.

With all his iconoclastic notions about furniture, Bennett defends the traditional distinction between craft and art, and abhors attempts to mix the two. "Some guys use a chair as a metaphor. They put it on the wall and people say, 'Oh, that's art!' Well, you can't sit on those chairs but you can get more money for them. I'm old-fashioned. My concern is first making a piece function, then making it as exciting to the eye as possible." Furniture provides only a context for Bennett's work. He proves that the breadth of one's imagination is the only limitation on creativity.

Bennett combined Finn-birch-plywood and gold leaf for this bench (1982), another in the series of benches with the same form—all done weeks apart (see p. 134).

JERE OSGOOD

ere Osgood is most widely known for his expressive yet subtle use of such modern woodworking techniques as tapered lamination and compound bending. Most pieces of Osgood's furniture resemble their conventional counterparts—chests look like chests, chairs look like chairs. But Osgood gives them a contemporary twist. The sides of an otherwise plain cabinet bulge slightly for an air of understated elegance, or the legs of a desk swerve sharply to one side, creating a sense of tension and motion. Every part has a quiet energy.

An early graduate of the School for American Craftsmen in Rochester, New York, Osgood inherited a respect for traditional methods from his teacher Tage Frid. But as he labored to create ever more sophisticated curved forms, these methods proved to be inadequate, so Osgood developed his own techniques. Says Osgood, "The techniques that I developed grew out of a searching in the design stage for new forms that would be strong and have a sense of freedom."

Osgood began experimenting with tapered laminations, where the form curves and tapers at the same time, in 1970, while searching for a sturdier method of constructing the legs of his Elliptical Shell desk. "I first made them by the usual process of building a lamination and cutting out the shape on a bandsaw," he says. "But when you cut into a lamination you weaken it and get those unsightly gluelines. So I tapered the laminates." Osgood believes this desk represents the first use of this technique. The shape of the shell forms a perfect ellipse—the front half rolls into the back with just ⅜-inch of clearance—and using a contrasting-colored wood for the inner laminae of the legs adds a distinctive touch.

Tapered lamination reappeared in the rail of a teak and ash desk Osgood built in 1982. (The desk marked the end of a two-year period of dwindling furniture production—a painful divorce and his appointment as acting director of Boston University's Program in Artisanry detracted from his studio time.) The sweep of the legs and shape of the top of the desk give it a sense of tension, as though it's being bent by the wind. "I wanted to establish a feeling of motion," Osgood says, "so it would appear to rotate as you walked around it." This desk was a model for a series of pieces making dramatic use of curved, tapering legs. A desk in the series that was constructed in 1985 has legs that run together and flow

The shell of this walnut desk (1970) forms a perfect ellipse. If Osgood were to make it again, he says he would draw the shell free-hand to give it a more human quality. The legs were his first use of tapered laminations.

JERE OSGOOD

With its curved legs piercing the top, this seven-legged coffee table (1984) was the precursor of the massive rosewood writing desk Osgood made the following year.

up through the writing surface to support a curved carcase (see p. 150).

Both desks illustrate Osgood's tendency to develop designs on paper and to use techniques that require careful forethought. This style of working often distinguishes college-trained woodworkers from self-taught craftsmen such as Wharton Esherick and Garry Knox Bennett, who design more intuitively and allow a piece to unfold as they work.

In 1969, Osgood created a chest of drawers using compound bent-stave lamination, a process he developed to form thin wooden shells curved in two planes. The gentle curves of the front and sides of the chest accentuate, rather than conflict with, the dramatic grain of the fiddleback Honduras mahogany. This chest was made for the Johnson Wax collection, and toured in the "Objects: USA" exhibition (p. 159).

Osgood also used compound bent-stave lamination in his Chest of Chair, made in 1972 (p. 155). The freeform shape of the drawer fronts and seemingly erratic placement of the handles contrast with the carefully matched veneer and precise curve of the seat.

Paradoxically, as Osgood developed

Every surface of this bubinga bench (1969) curves in a different direction.

JERE OSGOOD

greater fluency in creating new forms, his designs became more restrained. He minimized the curves of a chest of drawers he made in 1983 for "Bentwood Today," an exhibition of contemporary furniture sponsored by the Rhode Island School of Design (p. 154). "I wanted to simplify the form and play down the technique," Osgood says. "The slight curves and bulges of the cabinet are an attempt to return to the organic origins of the wood—its treeness—which flat boards do not express." For despite Osgood's interest in manipulating wood, he is committed to natural, treelike forms. But instead of incorporating slabs or branches into his furniture, he suggests these images through his elaborate techniques. Osgood defends his methods as consistent with the nature of the mate- Wanting to begin a new direction in his work with a dramatic piece, Osgood created this large writing desk in 1982. The legs are tapered laminations of ash with two thin layers of walnut in the middle. The solid-teak top is also curved to add to the feeling of strain and movement.

rial: "I feel that lamination the way I use it follows the growth patterns in a tree better than can be achieved with traditional joinery techniques using square milled-to-thickness lumber."

In contrast to the chest of drawers made in 1983, technique completely dominates a seven-legged coffee table Osgood made for "Art for the Table," a 1983 show at the American Craft Museum (p. 146). Here the treeness of the piece is achieved through rootlike legs that join into a stem and pierce the table's surface. Osgood calls the table "an exercise in repetition." Originally designed with three legs, Osgood gradually increased the number to seven because he liked the echoing effect.

Osgood began developing his methods around 1969, during the time Wendell Castle's sculptural stack-laminated pieces were in vogue. "I was at the opposite end of the design spec-

trum," he says. "While many woodworkers were making their furniture heavier, mine grew increasingly lighter." In addition, his work philosophy differs from that of a sculpture-oriented craftsman. "I place more emphasis on pre-planning than shaping. The form comes from bending the wood into a light shell instead of removing stock. I'd rather spend my time drawing and drafting and making jigs than chopping away lumber."

A major attraction of Osgood's methods is that they lend themselves to limited production runs. Although the time required to make a single piece is the same whether you carve solid stock to create the form or bend thin layers of wood, with lamination the craftsman is left with molds and jigs that can be used to create an identical piece in a fraction of the original time. The sculptor must perform the same processes

and invest the same time as for the first piece. Osgood discusses this benefit in his classes and in a series of articles he wrote for *Fine Woodworking* magazine, yet he rarely takes advantage of it, preferring instead to move on to new challenges.

Osgood has developed a considerable reputation as a chairmaker: His first chairs display a Danish influence; after returning from a year of study in Denmark in 1961, he made an easy chair that was bulky in appearance but simple in design (p. 157). In a captain's chair, strips of ash laminated into the back rail and legs give the chair a classic look of inlay (p. 159).

While exploring the relationship between joinery and design in 1965, Osgood developed the 7% chair (p. 153). It was an experiment in designing a chair that was light—in weight and appearance—and suitably strong. Ten

JERE OSGOOD

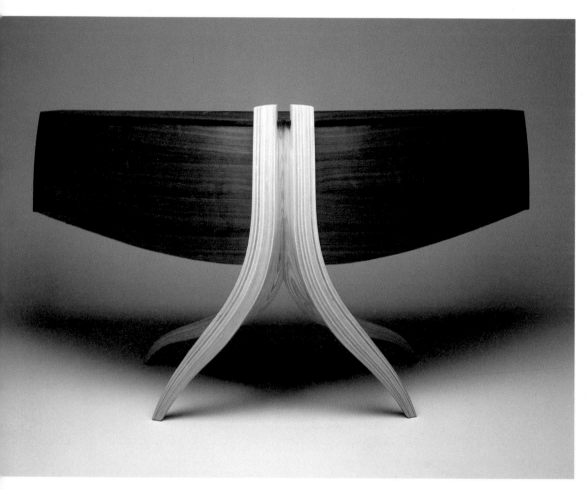

The culmination of a series of works with double-curved legs, Osgood used both double-tapered lamination in the ash legs and compound bent-stave lamination in the Brazilian-rosewood case of this desk (1985). The top is lined with black leather.

This staircase for the Balbirne home made in 1967 is completely suspended above the floor by long slats of cherry that hang from the walls and floor joists.

Originally intended as a study in joinery, the 7% chair (1965) later inspired Osgood to assign his students to make a chair as light as possible. The name refers to the requirement that the chair be seven percent stronger than its breaking point.

years later, he turned the problem into an exercise for his students by asking them to make a chair as light as possible without compromising structure. "When you're using heavier stock, the proportions and joinery don't matter as much because there's room for error. But by making everything light, you must have a firm grasp of how to design for joinery." The assignment became known as the 7% *problem* because Osgood would tell the students that the chair should be only seven percent stronger than its breaking point—a mythical point, impossible to calculate.

In the same vein as the 7% chair, Osgood made another, stronger desk chair (p. 158). A simple design, the chair contains just enough handwood to make it interesting. Its unusual strength comes from the way. the parts lock together. The side rails are dovetailed into the front legs and tenons of the front rail hold them in place. Three tenons fit together at the back, taking advantage of every possible gluing surface. Osgood has made six of these chairs—one in Andaman padauk was purchased by the Museum of Fine Arts in Boston. (Osgood's work also appears in the Rochester Institute of Technology and the American Craft Museum.)

Like Tage Frid, Osgood occasionally accepts commissions that simply offer an opportunity to produce good cabinetry. "They don't advance me as much as my one-of-a-kind furniture," he says, "but they provide a good income and keep my skills sharp." Robert Bruce Balbirne commissioned Osgood in 1967 and 1968 to create much of the interior woodwork and furniture for his house in Radnor, Pennsylvania, which was designed by famed architect I.M. Pei. Osgood constructed cabinets for the kitchen and baths as well as the outer doors of the house, two 6½-foot tall tamboured doors enclosing a bar, and the main staircase. The staircase is completely suspended and never touches the floor—the steps attach to long planks of cherry, elliptical in cross section, that hang from the walls and ceiling joists. Osgood believes a craftsman needs a solid background in construction methods. "We require our students to have a thorough understanding of basic technique. We hope it will become instinctive so they will naturally incorporate it into their thinking when they design."

Osgood was born on Staten Island, New York, in 1936. His father had a large shop in the basement of the

JERE OSGOOD

As Osgood developed greater proficiency with compound-bending techniques, his use of the process became more restrained and mature as in this chest of drawers (1983) of Andaman padauk with ebony pulls.

Osgood's playful Chest of Chair (1972) has a slightly bulging front. The wood is curly maple.

JERE OSGOOD

house, where young Osgood had his own workbench and learned to handle tools and machines. While in high school, he ran a part-time business from the shop, repairing furniture for neighbors.

He studied architecture at the University of Illinois from 1955 to 1957, but found he preferred interior detailing to designing buildings as a whole. In addition to the drafting skills Osgood acquired—"I'm not a natural drawer," he admits—he feels architecture provided him with a structured process for solving design problems on paper. After two years in Illinois, he transferred to the School for American Craftsmen on the advice of his father, who urged him not to compromise himself by pursuing a career he wouldn't enjoy. Osgood was already making a few furniture pieces and producing a line of wooden accessories, which he sold through a craft gallery. But he was told to drop his outside business by his teachers to concentrate on his studies. In desperate need of money to fund his education, Osgood used to sneak back

Osgood first applied the process of compound bent-stave lamination to this chest of drawers (1969) of fiddleback Honduras mahogany, which he created for the "Objects: USA" exhibition.

Upon returning from a year of study in Denmark in 1961, Osgood created this easy chair of curly maple and blue fabric. Although clumsy compared to his later chairs, it showed Osgood's tendency toward straightforward designs free of unnecessary embellishment.

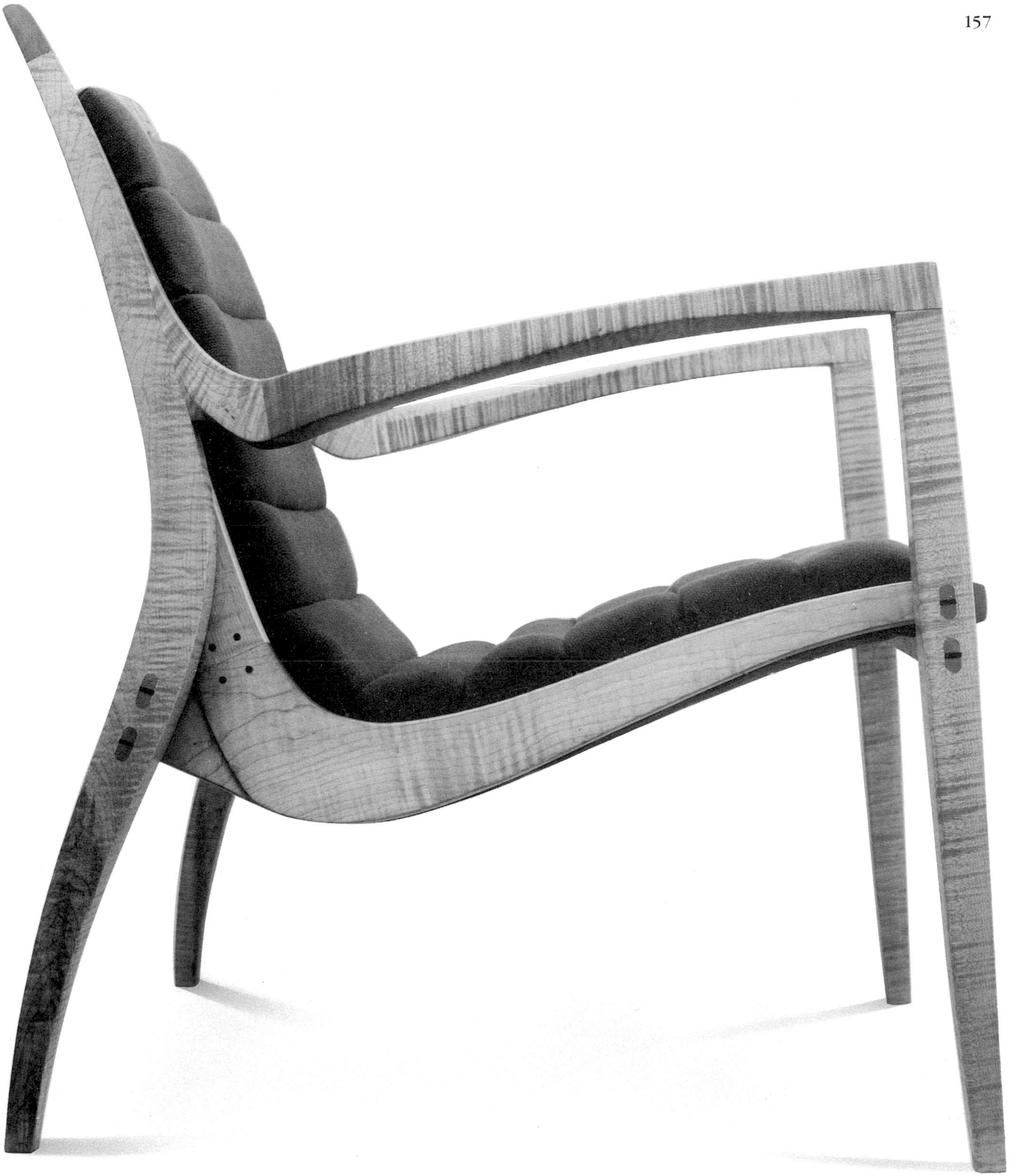

JERE OSGOOD

into the school shop at night and run the equipment with the lights off. On issues of design and technique, Osgood remembers himself as a somewhat obstinate student, often arguing with his teacher Tage Frid. But by the end of his training, he developed immense respect for Frid's personal energy and technical perception.

Immediately after graduating in 1960, Osgood traveled to Denmark to study furniture for a year through an educational organization known as the Scandinavian Seminar. "I learned a lot about how to approach materials and their use in furniture," he says, "and about the Danish approach to living as well." When he returned to the States, he established a shop in Connecticut, which served as his permanent home while he taught in various capacities from 1962 to 1974 at the Craft Students League in New York, the Brookfield Craft Center in Connecticut, the Philadelphia College of Art and the School for American Craftsmen. In

1975, he accepted a position along with his close friend Daniel Jackson to develop the wood program at the Program in Artisanry at Boston University, which had begun a few months earlier under James Krenov's brief incumbency. "What excites me about teaching," Osgood says, "is working with talented students who you can nurture over several years." The Program moved in 1985 to the Swain College of Design in New Bedford, Massachusetts, where Osgood continues to teach part-time

This desk chair is one of the few pieces in which Osgood makes use of hand detailing. This version in Andaman padauk and black leather (1982-83) was commissioned by the Boston Museum of Fine Arts for their permanent collection.

while devoting most of his efforts to making furniture.

As a child, Osgood's family spent their summers in Vermont and the experience left a deep impression on him. "I saw there was a type of living other than that in New York City—the total change of pace and peace amazed me even as a young boy," he recalls. Almost 40 years later, he returned to rural New England to make his home in Wilton, New Hampshire. His small house connects to a spacious, well-organized workshop where the walls are lined with jigs and forms from past work. Osgood has two sons, Leif and Mark. Like his father, Osgood taught his children to use tools—Mark, the youngest, is entirely at home in the shop and since age nine has demonstrated particular skill on the wood lathe.

Osgood represents the new generation of university-trained woodworkers and the application of modern techniques at their best. Rather than allow himself to be carried away by the new ability to force wood into previously unattainable forms, he uses it to support his design ideas. As his methods become more sophisticated, the designs become more mature and thoughtful. Despite his innovative methods, Osgood is reluctant to embrace some of the new trends in the craft, like painting furniture or using other materials. "I'm a long way from exhausting the potential of wood," he says.

The laminations were elevated to design elements in this walnut desk chair (1969) where Osgood sandwiched thin strips of ash into the back rail and back legs.

SELECTED BIBLIOGRAPHY

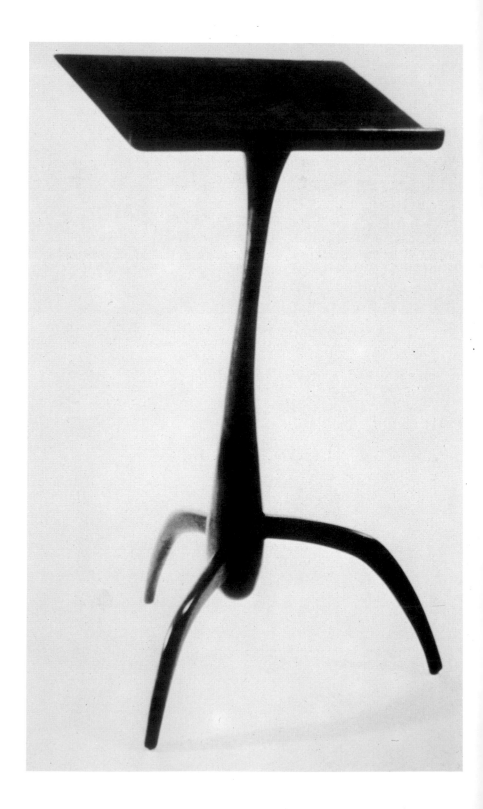

The legs of Arthur Espenet
Carpenter's walnut music stand
(1970) have been likened to a
crawling insect since they meet the
stalk at different heights.

WHARTON ESHERICK

Benson, Gertrude. "Wharton Esherick." *Craft Horizons*, January/February 1959, pp. 33-37.

Grafly, Dorothy. "Wharton Esherick." *Magazine of Art*, January 1950, pp. 9-11.

Hoffman, Marilyn. "Individuality Wins." *The Christian Science Monitor*, December 31, 1958, p. 6.

Kelsey, John. "Wharton Esherick." *Fine Woodworking*, Summer 1977, p. 45.

Levins, Hoag. "A Shy Artist's Museum of Design." *Philadelphia Inquirer— Today*, May 26, 1974, pp. 24-26, 28.

Maloof, Sam and Wendell Castle. "Wharton Esherick 1887-1970." *Craft Horizons*, August 1970, pp. 10-17.

McKee, Tally. "Sculpture in Architecture." *Charette*, December 1952, pp. 11-14.

Museum of Contemporary Crafts. *The Furniture and Sculpture of Wharton Esherick*. Catalog of exhibit: December 12, 1958-February 15, 1959. New York: Museum of Contemporary Crafts of the American Craftsmen's Council, 1958.

Page, Mariam. "The sculptural environment of Wharton Esherick." *Interiors*, February 1959, pp. 92-95.

Rochberg, Gene. *Drawings by Wharton Esherick*. New York: Van Nostrand Reinhold, 1978.

Stone, Michael. "Wharton Esherick." *Fine Woodworking*, November/December 1979, pp. 50-57.

The Wharton Esherick Museum. *The Wharton Esherick Museum: Studio and Collection*. Catalog of collection. Paoli, Pa.: Wharton Esherick Museum, 1977.

Yarnall, Sophia. "Sculptural Wood Creates the Unique Interiors of the Curtis Bok House." *County Life and The Sportsman*, June 1938, pp. 67-74.

GEORGE NAKASHIMA

"The Craftsman." *Life*, June 12, 1970, pp. 75-78.

"Craftsmanship in Architecture." *Craft Horizons*, May/June 1956, pp. 26-31.

George Nakashima, Woodworker. Catalog of work. New Hope, Pa.: George Nakashima.

Hepler, Paul H. *A Study of the Work Training and Craft Processes of a Contemporary American Wood Craftsman*. Ed.D. thesis, Teachers College, Columbia University, 1969.

Kaufmann, Edgar, Jr. "Nakashima, American Craftsman." *Art in America*, December 1955, pp. 30-33.

Kelsey, John. "George Nakashima." *Fine Woodworking*, January/February 1979, pp. 40-46.

Lyon, Mary. "Nakashima." *Craft Horizons*, Autumn 1949, pp. 16-19.

Nakashima, George. "Actuality." *Perspecta*, March 1955, pp. 26-33.

Nakashima, George. "Credo, Moral aspects of design." *Liturgical Arts*, August 1964, pp. 128-129.

Nakashima, George. *The Soul of a Tree*. New York: Kodansha, 1981.

Reif, Rita. "Craftsman's Feeling for Woods Lends Greatness to His Furniture." *The New York Times*, August 25, 1958, p. 25.

BOB STOCKSDALE

Anderson, Chuck. "Tools Need Not Be Brand New For a Fine Shop." *Oakland Tribune*, November 7, 1976, pp. 8-H, 10-H.

Huff, Darrell. "Bowls Are His Business." *Popular Science Monthly*, August 1956, pp. 165-168.

Kelsey, John. "The Turned Bowl." *Fine Woodworking*, January/February 1982, pp. 54-60.

Museum of Contemporary Crafts. *Wood Turnings by Bob Stocksdale*. Catalog of exhibit: January 25-March 16, 1969. New York: Museum of Contemporary Crafts of the American Craftsmen's Council, 1969.

Newhall, Barbara Falconer. "Woodcrafting, Bay Area Style." *San Francisco Chronicle*, December 21, 1978, p. 24.

Stocksdale, Bob. "Exotic Woods." *Fine Woodworking*, Fall 1976, pp. 28-32.

"Wooden Tableware." *Science Illustrated*, February 1949, pp. 78-79.

TAGE FRID

Davidson, Margaret. "Unforgettable Kitchens." *Ladies' Home Journal*, March 1960, pp. 64-73.

Frid, Tage. *Tage Frid Teaches Woodworking: Book 1— Joinery*. Newtown, Conn.: Taunton Press, 1979.

Frid, Tage. *Tage Frid Teaches Woodworking: Book 2—Shaping, Veneering, Finishing*. Newtown, Conn.: Taunton Press, 1981.

Frid, Tage. *Tage Frid Teaches Woodworking: Book 3— Furniture*. Newtown, Conn.: Taunton Press, 1985.

Kelsey, John. "Tage Frid." *Fine Woodworking*, May/June 1985, pp. 66-71.

Murkowski, Carol. "Woodworker lends life to his creations." *The Anchorage Times*, May 8, 1983, pp. M-1,2.

Rowley, Gordon E. "Tage Frid: Dean of American Woodworking." *The Observer—North County*, March 15, 1984, pp. 1-2.

SAM MALOOF

Loney, Glenn. "Sam Maloof." *Craft Horizons*, August 1971, pp. 16-19.

Maloof, Sam. *Sam Maloof, Woodworker*. New York: Kodansha, 1983.

Mastelli, Rick. "Sam Maloof." *Fine Woodworking*, November/December 1980, pp. 48-55.

Masterpieces from the Boston Museum. Boston: Museum of Fine Arts, 1981, p. 95.

Webster, John. "Handsome Furniture You Can Build." *Better Homes and Gardens*, March 1951, pp. 258-259.

ARTHUR ESPENET CARPENTER

Carpenter, Arthur Espenet. "The Chairs of Charles Eames." *Craft Horizons*, April 1973, pp. 20-21, 55.

Carpenter, Arthur Espenet. "The Rise of Artiture." *Fine Woodworking*, January/February 1983, pp. 98-103.

"A Fine Round of Woodwork." *Life*, December 9, 1966, pp. 140-144, 147.

Mastelli, Rick. "Art Carpenter." *Fine Woodworking*, November/December 1982, pp. 62-68.

Stone, Michael. "The Espenet Style." *American Craft*, June/July 1982, pp. 6-9.

JAMES KRENOV

Arwidson, Bertil. "James Krenov—inspired by the Wood." *Swedish Interior*, No. 2, 1978, p. 50.

Krenov, James. *A Cabinetmaker's Notebook*. New York: Van Nostrand Reinhold, 1976.

Krenov, James. *The Fine Art of Cabinetmaking*. New York: Van Nostrand Reinhold, 1977.

Krenov, James. *The Impractical Cabinetmaker*. New York: Van Nostrand Reinhold, 1979.

Krenov, James. *James Krenov, Worker in Wood*. New York: Van Nostrand Reinhold, 1981.

Krenov, James. "Wood." *Craft International*, January 1983, pp. 18-19.

Krenov, James. "Wood: ' . . . the friendly mystery . . . ' " *Craft Horizons*, March/April 1967, pp. 28-29, 54.

Stone, Michael. "The Quiet Object in Unquiet Times." *American Craft*, February/March 1984, pp. 39-42.

WENDELL CASTLE

Castle, Wendell. "Design Considerations." *Fine Woodworking*, Winter 1976, p. 27.

Castle, Wendell and David Edman. *The Wendell Castle Book of Wood Lamination*. New York: Van Nostrand Reinhold, 1980.

Chastain-Chapman, A. U. "Wendell Castle Tries Elegance." *Fine Woodworking*, September/October 1983, pp. 68-73.

Hemphill, Christopher. "Against the Grain: The Art of Wendell Castle." *Town & Country*, May 1984, pp. 243, 248, 250, 252.

Kelsey, John. "Stacking." *Fine Woodworking*, Winter 1976, pp. 22-26.

Wilson, Tom Muir. *The Music Rack*. New York: American Craft Council, 1964.

GARY KNOX BENNETT

Johnson, Beverly Edna. "Blunk, Bennett, Bauer & Barbara: The Four Bs are Designing People." *Los Angeles Times—Home*, November 18, 1973, pp. 14-19.

Kelsey, John. "Portfolio: Garry Knox Bennett." *Fine Woodworking*, March/April 1984, pp. 79-81.

Levine, Melinda. "Garry Bennett's Sculptural Furniture." *Artweek*, August 11, 1979, p. 3.

Marlowe, John. "With the Wind and Stairs in his Heirs." *WestArt Annual*, December 3, 1976, pp. 10-13.

Stone, Michael. "Garry Knox Bennett." *American Craft*, October/November 1984, pp. 22-26.

JERE OSGOOD

Osgood, Jere. "Bending Compound Curves." *Fine Woodworking*, July/August 1979, pp. 57-60.

Osgood, Jere. "Bending a Tray." *Fine Woodworking*, Summer 1977, pp. 62-64.

Osgood, Jere. "Bent Laminations." *Fine Woodworking*, Spring 1977, pp. 35-38.

Osgood, Jere. "Tapered Lamination." *Fine Woodworking*, January/February 1979, pp. 48-51.

Somerson, Rosanne. "Perfect Sweep." *American Craft*, June/July 1985, pp. 30-34.

GENERAL

The Akron Art Institute. *Why Is An Object*. Catalog of exhibit: September 11-November 4, 1962. Akron, Ohio: The Akron Art Institute, 1962.

Buffalo Craftsmen and the Burchfield Center. *Language of Wood*. Catalog of exhibit: October 19-November 30, 1975. Buffalo, N.Y.: Charles Burchfield Center, 1975.

Fairbanks, Jonathan L. and Elizabeth Bidwell Bates. *American Furniture, 1620 to the Present*. New York: Marek, 1981.

Garner, Philippe. *Twentieth-Century Furniture*. New York: Van Nostrand Reinhold, 1980.

Makepeace, John. *The Art of Making Furniture*. New York: Sterling, 1981.

"The New American Craftsman: First Generation." *Craft Horizons*, May/June 1966, pp. 15-19.

Nordness, Lee. *Objects: USA*. New York: Viking, 1970.

Smithsonian Institution and Minnesota Museum of Art. *Woodenworks*. Catalog of exhibit: January 28-July 19, 1972. Renwick Gallery of the National Collection of Fine Arts, Smithsonian Institution, Washington, D.C. and October 12-December 31, 1972, Minnesota Museum of Art, Saint Paul, Minnesota, 1972.

Wallance, Don. *Shaping America's Products*. New York: Reinhold, 1956.

Castle's Pinwheel game table represents a transition from his stacked and *trompe l'oeil* furniture to his later interest in historic furniture styles. This table (1980) is walnut with a spalted curly maple top.

PHOTOGRAPH CREDITS